MOBILE LEARNING AND HIGHER EDUCATION

Mobile Learning and Higher Education provides case studies of mobile learning in higher education settings to showcase how devices can transform learning at the undergraduate and graduate levels. With the rapid diffusion of networked technologies among the adult populations of many countries and the supersession of the once-ubiquitous lecture approach with active learner-centered teaching for deep understanding, mobile devices are increasingly used in higher education classrooms to offer unique and effective new approaches to teaching and learning. A cutting-edge research volume, this collection also provides a springboard for building better practices in higher education institutions.

Helen Crompton is Assistant Professor of Instructional Technology in the Department of Teaching and Learning at Old Dominion University, USA.

John Traxler is Professor of Mobile Learning and Director of the Institute of Education at the University of Wolverhampton, UK.

MOBILE LEARNING AND HIGHER EDUCATION

Challenges in Context

Edited by Helen Crompton and John Traxler

Routledge
Taylor & Francis Group
NEW YORK AND LONDON

First published 2018
by Routledge
711 Third Avenue, New York, NY 10017

and by Routledge
2 Park Square, Milton Park, Abingdon, Oxon, OX14 4RN

Routledge is an imprint of the Taylor & Francis Group, an informa business

Library of Congress Cataloging-in-Publication Data
A catalog record for this book has been requested

ISBN: 978-1-138-23876-3 (hbk)
ISBN: 978-1-138-23877-0 (pbk)
ISBN: 978-1-315-29673-9 (ebk)

Typeset in Bembo
by Deanta Global Publishing Services, Chennai, India

CONTENTS

CONTRIBUTORS

Frances M. Boylan is Head of eLearning Support & Development (Acting) at the Dublin Institute of Technology in Dublin, Ireland. Dr Boylan has an interest in the integration of mobile apps into learning, teaching and assessment practices, and has both initiated and facilitated many continuing professional development opportunities for educators in the area.

Cheryl Brown is Senior Lecturer and part of the Learning Technologies Team within the Centre for Innovation in Learning and Teaching (CILT). Her main research focus over the last few years has been around the role personal mobile devices play in students' learning in a developing context and the development of students' digital literacy practices. Cheryl is also the Principal Investigator on the Department of Higher Education and Training/University of Cape Town (UCT) Personal Mobile Device Project in Cape Town, South Africa.

Cathy Cavanaugh is Head of Learning and Leadership in Digital Transformation at Catholic Education Western Australia. Cathy's academic research and publications focus on teaching and learning in virtual schools, online and blended learning, teacher development, and mobile learning. She has consulted with governments and organisations around the world on education programs and policy.

Thomas D. Cochrane is an academic advisor in eLearning and learning technologies at Auckland University of Technology in Auckland, New Zealand. Thomas is widely published in the areas of mobile learning and pedagogical change.

Helen Crompton is a Professor of Instructional Technology at Old Dominion University in Norfolk, Virginia USA. She is a highly experienced researcher, educator, author and presenter in the field of educational technology. She draws from over 20 years in education and a PhD in educational technology and mathematics education from the University of North Carolina at Chapel Hill. Dr. Crompton has consulted for governments and two divisions of the United Nations; UNESCO the intelligence division and ITU the technology division.

Mark M. Diacopoulos is a PhD Candidate in Curriculum and Instruction at Old Dominion University in Norfolk, Virginia USA. His research focus is how technology can be integrated into social studies education. Mark works as an Instructional Specialist in a large school system in southeast Virginia.

Jim G. Ffrench is a Lecturer and has carried out this research as part of an MSc in eLearning, with final year apprentices in refrigeration and air conditioning in the School of Mechanical and Design Engineering in Dublin Institute of Technology, Ireland.

Derek France is Professor of Pedagogy in Geographical Sciences at the University of Chester in Chester, England. He was awarded a National Teaching Fellowship in 2008 and is a coeditor for the *Journal of Geography in Higher Education*. His research interests cover technology enhanced learning and the development of research in fieldwork and mobile technologies.

Michael S. Gallagher is a Research Associate at the Centre for Research in Digital Education at the University of Edinburgh in Edinburgh, UK and Director of Panoply Digital, a consultancy dedicated to mobile for development (M4D). His research focus is on mapping the effects of mobility and mobile technology on built environments and urban spaces to support practice in the humanities in higher education, particularly in the Asia Pacific region.

Ian Glover is Senior Lecturer in Academic Development, with an emphasis on Digital Capability and Technology Enhanced Learning, at Sheffield Hallam University in Sheffield, England. He has held a variety of Educational Technology roles at different UK universities and has a PhD related to the development of collaborative online learning tools.

Andrew T. Greenwood is an officer in the US Coast Guard currently assigned to FORCECOM as a Training Manager. As a recent graduate from San Diego State University's Learning Design and Technology program, his studies focused on mobile learning, augmented reality and its application in formal and informal learning environments. His research focused on learning theory and performance analysis.

Jace Hargis is Director of the Center for Engaged Teaching at the University of California, San Diego. His research agenda focuses on how people learn with the use of appropriate, relevant and meaningful instructional technologies.

Genevieve Haupt joined the Learning Technologies Team within the Centre for Innovation in Learning and Teaching (CILT) in September 2015 as the project manager and researcher on the Department of Higher Education and Training/ University of Cape Town (UCT) Personal Mobile Device Project in Cape Town, South Africa. Prior to her appointment at CILT she worked in the Education and Skills Development Unit at the Human Sciences Research Council where she played various researcher roles in projects across the field of higher education.

John Hedberg is the Millennium Innovations Chair of Information and Communication Technologies (ICT) and Education in the School of Education at Macquarie University in Sydney, Australia.

Billie Jean Holubz is Assistant Professor of Education for Liberty University Online in Lynchburg, VA, USA, Doctoral Chair for Grand Canyon University in Phoenix, Arizona, USA, and Adjunct Instructor for City University in Seattle, Washington, USA. With over 25 years in K-12 education, she is a proponent of instructional technology, serves as a leader in the implementation of mobile technology, and is associated with the International Society of Technology's Mobile Learning Network (ISTE MLN). Dr Holubz is an action researcher and author, and incorporates mobile learning as an integral part of student learning beyond the classroom. Passion for mobile learning, extensive experience, and research practices equip this researcher as an exceptional leader in educational technology and mobile learning.

Aditi Hunma is a lecturer in the Centre for Higher Education Development at the University of Cape Town in Cape Town, South Africa. She teaches on introductory academic literacy courses at the undergraduate level, designs online language modules, and facilitates online and face-to-face research writing interventions at the postgraduate level. These interventions are framed within academic development and offered under the banner of social redress and transformation. Her research interests revolve around looking at online spaces as transformative spaces.

Katharine E. Jewitt is a researcher in virtual reality at the School of Education, University of Glasgow in Scotland and Visiting Honorary Associate and Lecturer in the Open University's Faculty for Science Technology Engineering and Mathematics in the United Kingdom. Katharine works as an educational technology consultant and researches technology enhanced learning in mobile environments.

Martin P. Khechara is a senior lecturer in Biomedical Science at the University of Wolverhampton in Wolverhampton, England. He has been using video for learning since 2009 and actively researches the use of technology and the flipped classroom for teaching in the healthcare sciences.

Alice L. Mauchline is Senior Research Fellow at the University of Reading in Reading, England. She has research interests in sustainable crop production approaches. She also researches innovative fieldwork pedagogies with the Enhancing Fieldwork Learning project and has published on a range of fieldwork and active learning topics.

Danielle L. Mulrennan is a Lecturer in Journalism at Auckland University of Technology, in Auckland, New Zealand. Her research interests are within journalism studies where she is particularly interested in the application of mobile social media within journalism education, and the relationship between the culturally deaf and the television broadcast news media.

Julian Park is Head of the School of Agriculture, Policy and Development at Reading University in Reading, England. He has played a pivotal role in bioscience education in conjunction with the Higher Education Academy. He was awarded a National Teaching Fellowship in 2008 and is a Principal Fellow of the Higher Education Academy.

Kumaran Rajaram is Senior Lecturer in the Division of Strategy, Management, and Organization in the College of Business at Nanyang Technological University, Singapore.

Helen Rodger is Senior Lecturer in Academic Development, with an emphasis on Digital Capability and Technology Enhanced Learning, at Sheffield Hallam University in Sheffield, England. Her current research is around the use of social media for learning in higher education, and understanding the role of technology in the delivery of a flexible curriculum.

Barry J. Ryan is Lecturer in Biochemistry and Biotechnology at Dublin Institute of Technology in Dublin, Ireland. Dr. Ryan has a deep interest in enhancing learning through judicious technology integration, and research informed adoption, in all learning environments.

Thulile P. Shandu is a lecturer in the Department of English Studies at the University of South Africa (UNISA) in Pretoria, South Africa. She is currently a doctoral candidate with UNISA in the area of Mobile Learning in Open Distance Learning. Her other research interests are in Applied English Language Studies as well as in language and power in multilingual contexts.

Helen A. Sissons is a senior lecturer in Journalism at Auckland University of Technology, in Auckland, New Zealand. Helen spent 17 years as a journalist in the US and the UK, the last ten with the BBC as a television journalist and reporter. She is the author of *Practical Journalism: How to write news*.

Sara C. Smith is Senior Lecturer in Biomedical Science at the University of Wolverhampton in Wolverhampton, England and a registered biomedical scientist. Her research focuses upon the development of practitioner capability and the application of innovative teaching approaches.

Michael Stevenson is a Postdoctoral Research Fellow in the Department of Educational Studies at Macquarie University in Sydney, Australia. His research explores the role of technology-enabled Personal Learning Networks in teacher professional learning.

John Traxler was Professor of Mobile Learning, the world's first, from September 2009, and now Research Professor of Digital Learning in the Institute of Education at the University of Wolverhampton in Wolverhampton, England. He is one of the pioneers of mobile learning, associated with mobile learning projects since 2001 when he was evaluator for m-learning, the first major EU project. He is a Founding Director and was Vice-President of the International Association for Mobile Learning, responsible for the annual international mLearn research conference running since 2002.

Minjuan Wang is Professor of Learning Design and Technology at San Diego State University in San Diego, California. She was a distinguished research Professor (adjunct) of Shanghai International Studies University in Shanghai, China. Her research specialties focus on the sociocultural aspects of online learning, mobile learning, and augmented reality. She is a chief editor for one of the European Alliance for Innovation (EAI) journals, and also serves on the editorial boards for four other education technology journals.

Katharine Welsh is a Senior Lecturer in physical geography at the University of Chester in Chester, England with interests in fluvial geomorphology. She is a Fellow of the Higher Education Academy and researches the benefits of using iPads and the concept of 'Bring Your Own Device' through involvement in the Enhancing Fieldwork Learning project.

W. Brian Whalley is a Visiting Professor at the University of Sheffield in Sheffield, England. He formerly researched and taught at the Queen's University of Belfast in Belfast, Northern Ireland where he was Professor of Geomorphology. Despite being 'retired' he still carries out field research in geomorphology and is developing pedagogic ideas using mobile devices. He was awarded a UK National Teaching Fellowship in 2008.

INTRODUCTION

Helen Crompton and John Traxler

Mobile learning (mlearning) is a potential game-changer in higher education (HE). In many countries, the adult population has more than one mobile device per person (Statista, 2016). With the rapid diffusion of mobile technologies, these devices are now in many HE classrooms and offering new approaches to learning. The lecture approach, which was common in many HE institutions across the world, is now slowly being replaced by learner-centered teaching (Wright, 2011). This approach transforms students from passive learners who only achieve surface-level learning to students actively involved in their learning, gaining a deep understanding of the concepts. Mobile devices offer new affordances to transform learning. However, this shift in learning is not simple and bigger issues need to be addressed, such as equity, scale, sustainability, and change management. The purpose of this book is to present a collection of theoretical foundational and empirical findings to address these bigger issues at the undergraduate and graduate level and provide a better understanding to research and practitioner communities of ways to successfully utilize mobile devices in the higher educational context.

Mobile Learning

To begin, the understanding should be clear on what mobile learning is and what it is not. In this book, the starting definition is that mobile learning is "learning across multiple contexts, through social and content interactions, using personal electronic devices" (Crompton, 2013, p. 4). This definition is highlighted as it encompasses the main assets and opportunities available when using mobile devices for learning. It highlights the multiple contexts, which refers to the open availability to learning in a variety of locations and that mobile learning can be used across any content. Social and content interactions refer to the mix of people and materials that the learner can utilize for learning. Finally, the definition includes the devices that provide the learner with those affordances. This definition also mirrors the various authors' voices in this book as it places the pedagogy before the device.

Transformative Pedagogies

As Traxler (2013) described, "Mobile technologies are the global context and education should reflect that context. Learning process will need to reflect that shift and so will teaching" (p. 247). The unfortunate reality is that these technologies "can also be used to support outmoded teaching and learning approaches" (NZMoE, 2012, p. 2). The pedagogical shift to contextualized, learner-centered, out-of-the-box learning is highlighted in the recent educator standards from the International Society for Technology in Education. The booklet to accompany the standards delineated the shift in trends with the 1998 standards in focusing on technology, changing to learning with technology with the 2007 standards, to now transforming learning in the 2017 standards (Crompton, 2017a). In the 2007 revision of the educator standards, it appeared that the focus was correctly moved to be on the pedagogy; however, the technology was merely used with dated pedagogies as these twenty-first century technologies were used for twentieth century teaching approaches. With the recent 2017 revision of the standards, educators are asked to transform learning and pedagogy using the technologies, to go beyond replicating past practices.

Chapters 1 to 7 in this book delineate ways learning is transformed with mobile devices. In Chapter 1, Whalley, Mauchline, France, Park, and Welsh summarize findings from a study involving mobile technologies in a fieldwork learning project. In this study, the researchers report that mobile devices could be used to facilitate innovative pedagogies and promote active learning in various field and out-of-class situations. For example, geology students were able to take angular measurements and measure the direction of rock strata while using the devices and collect this information on notes which also provided a geo-referenced location to where the measurements were taken.

In Chapter 2, Cochrane, Sissons, and Mulrennan conduct a curriculum redesign for a journalism course. This five-year journey takes them through initial reactions, to mobile social media, and then to the curriculum redesign to integrate mobile learning. To enable the students to develop expertise in the use of social and mobile tools in an authentic and self-sustaining manner, the researchers adopt a student-determined learning approach (heutogogy) on student-owned mobile devices to enable learning to move outside the classroom and continue beyond the university. From this study, the researchers extracted and presented a set of transferable design principles to be used in other disciplines.

In Chapter 3, Khechara and Smith examine the flipped classroom approach with mobile devices. The authors presented the findings from three different studies at both undergraduate and graduate levels in biomedical science education. The studies were grounded in a social constructivist paradigm with learning as a collaborative event. The findings of these studies organized by academic level reveal the changes in attitudes and emotional state of students as they engaged in the flipped learning experiences.

In continuing the exploration of innovative pedagogies, Chapters 4 and 5 present studies involving augmented and virtual reality. In Chapter 4, Greenwood and Wang introduce a theoretical foundation for implementing augmented reality in education and address key issues of how mobile learning can break out of small niches into wider institutional acceptance. In this chapter, the authors initially provide a snapshot of various theoretical frameworks, such as immersive learning theory, connectivism, social interaction, social development theory, More Knowledgeable Other (MKO), and Zone of Proximal Development (ZPD). They then move on to an overview of augmented reality in education and a discussion as to the wider institutional acceptance involving challenges, barriers, organizational culture, and regulation.

In Chapter 5, Jewitt focuses on virtual reality, which is seeping into the educational arena as another innovative pedagogy. Jewitt uses virtual worlds to provide collaborative spaces for learners to meet to avoid the students feeling anxious as they use role play for learning. Using the virtual reality environment, learners engage anonymously through the adoption of an avatar and learn from their peers through observation, discussion, and peer assessment.

Demonstrating the flexibility mobile devices provide to transform pedagogical practice, Shandu in Chapter 6 studies open distance learning. In this chapter, students attending the University of South Africa harness the ubiquity, versatility, and availability of mobile phones for student support. Framing the research within the Community of Inquiry, Social, Cognitive, and Teaching Presences, the authors examine how students use mobile devices to enhance their vocabulary learning. At the close of the chapter, Shandu provides a set of guidelines for the use of mobile technologies to enhance students' vocabulary.

Implementing transformative pedagogies requires instructors to rethink the way they may have taught for many years. In Chapter 7, Glover and Rodger make a case for the end of mobile learning as a discrete area and argue for mobile devices to be used as potential tools to be integrated into the wider learning ecosystem. They describe the tension between the way learners work and want to work on the one hand and how instructors want them to work on the other; learners organically create their own learning networks and experiences while instructors seek to create discrete, defined mobile learning experiences. Glover and Rodger summarize the current literature demarcating the shift towards more active pedagogies and societal and institutional changes.

As these pedagogical shifts take place, it is interesting to note both the HE students' and instructors' perspectives on mobile learning. Empirical evidence shows that the beliefs held by the individual on the efficacy of mobile learning will determine the way mobile devices are integrated into learning (Crompton, 2017b). In Chapter 8, Holubz uses a qualitative phenomenological approach to explore online students' perspectives of connectivity, navigation, and experiences with mobile devices. Similar to Chapter 5, the devices are used in an online environment. In this study, Holubz uses connectivism theory that views learning within the person. Navigation theory is also used as instructors become mentors and coaches that enable students to learn through many different means, locations, and connections to knowledge, rather than learning specific pre-selected information. From the student perceptions, Holubz presents a set of guiding points that can be used to inform instructors and administrators in the HE context in successfully implementing a mobile learning initiative.

In Chapter 9, Boylan uses a thematic analysis to take a qualitative look at perspectives of instructors who had completed a 12-day mobile learning program. The *12 Apps of Christmas* course was designed to raise awareness among instructors of the benefits mobile technologies and mobile apps present for learning, teaching, and assessment. For each of the 12 days, a page on a website was opened which contained information about a specific mobile application that the instructor could explore. Following the 12 days, Boylan gathered educators' perspectives on the integration of mobile applications, mobile literacy, the value of training, and barriers to learning with mobile devices. In conclusion, the findings are presented with a set of key considerations for those implementing mobile learning initiatives in HE. In this study, Boylan reports that within the Irish context, mobile learning is in the embryotic stage.

Cultural Implications

The geographical location can make a difference to mobile learning initiatives due to cultural norms, infrastructure, societal values, and trends in learning. In the various chapters of this book, the studies are based in many different geographical contexts. Figure I.1 shows the many locations of the world where these studies took place.

Chapters 10, 11, and 12 provide a diverse cultural window into mobile learning in South Africa, the United Arab Emirates, and South Korea. In Chapter 10, Brown, Haupt, and Hunma investigate models to provide equitable access to Bring Your Own Device initiatives for teaching and learning in a context of unequal access. The authors describe that although mobile learning initiatives are not new to HE in South Africa, it cannot be assumed that students have access to devices or that students have the digital skills to use the devices. Many students are "digital strangers", outsiders to the digital world. In this chapter, the authors focus on the influence of mobile devices on the teaching practice of instructors who have not had experience using educational technologies. Both the challenges and the educational opportunities are presented.

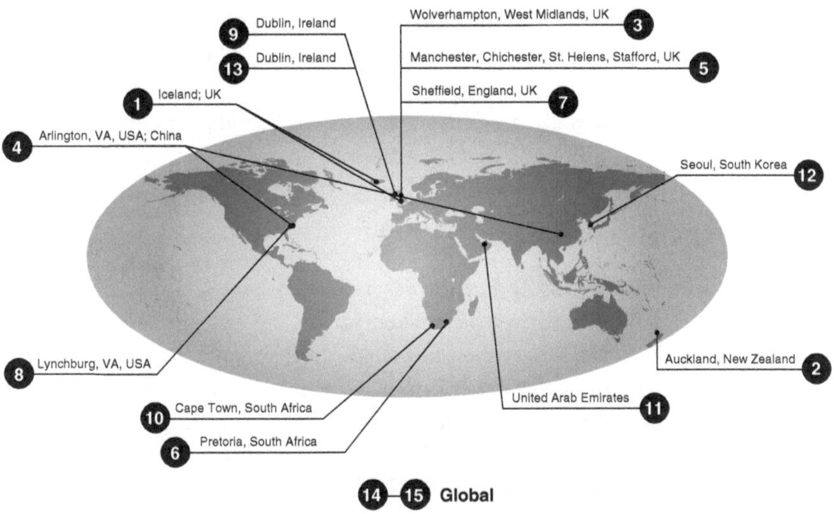

FIGURE I.1 The International Scope of This Book's Studies on Mobile Learning Initiatives

Chapter 11 moves to the other end of the spectrum for mobile learning resources as Cavanaugh and Hargis examine the mobile learning ecosystem in the United Arab Emirates with a program presented as the largest educational iPad initiative in the world. In this mobile learning initiative, the emphasis was placed on the pedagogy rather than the technology, in an effort to transform the learning environment to increase engagement and student-centered practice for the incoming students. Using the lens of organizational congruence, the authors present the experiences of students in blended courses, faculty instructors, and faculty leaders using a holistic data strategy. The findings illuminate the adoption and integration of new learning environments.

Gallagher elucidates the digital milieu of South Korea in Chapter 12. Gallagher describes the problematic tension between a top-down model of smart learning in the formal space and the sophisticated use of mobile technology and social media in the informal space, with use that directly affects the nature and structure of formal disciplinary participation. The tensions are explored as an overview of the state of mobile technology use in HE in South Korea is revealed. Gallagher shows how the use of mobile devices is governed by the informal mobile cultures and practices that have arisen amidst domestically produced mobile technology and mobile applications.

In many cultures of the world, learning through apprenticeships is a valued pedagogical method dating back many centuries. Two of the chapters in this book focus on apprenticeships with mobile learning. In Chapter 5, Jewitt writes about the use of apprenticeships in virtual reality environments. In Chapter 13, Ffrench and Ryan study apprenticeship learning with mobile devices in science. Ffrench and Ryan wanted to determine the efficacy of using bespoke mobile learning resources to improve student motivation. The authors begin this chapter by presenting the theories for motivation and mobile learning. Then—using an action research model of planning, acting, observing, and reflecting—they examine if mobile learning resources had a positive influence on learning. These authors also share various barriers to the use of these technologies.

The Global View

Each of these studies in this book provides a detailed look at a specific context where mobile learning initiatives have taken place across the globe. These valuable studies dig deep to uncover the finer details of that context to the academic and practitioner community. However, the research on mobile learning needs to also be examined on a larger scale to better understand trends and build a comprehensive picture of the field (Oakley, 2012; Sandelowski, Voils, Leeman, & Crandlee, 2011). Chapters 14 and 15 provide that global view on mobile learning through systematic reviews of the recent mobile learning literature. In Chapter 14, Diacopolis and Crompton reveal findings of a thematic review of recent research on social studies education. From an initial analysis of 2,086 studies, inclusion/exclusion criteria reduced the number to 11 studies on mobile learning in HE social studies. These studies were examined and three themes are presented.

In Chapter 15, Stevenson, Rajaram, and Hedberg conduct a systematic review on mlearning across all subjects in HE. The authors review the mobile learning studies to identify the nature of effective and ineffective learning in mobile-rich classrooms and investigate the cultural factors that shape learning with mobile devices within and between HE institutions. From the findings, the authors present key principles of effective learning in mobile-rich HE classrooms.

The studies in this book provide a guide to showcase examples of how mobile devices are being used to transform learning at the undergraduate and graduate level. These studies all use mobile learning as the driving force toward unique and effective HE teaching and learning. More importantly, the studies in this book address the bigger issues of equity, scale, sustainability, and change management with specific guidelines and principles uncovered from the research that can be used as a springboard for building better practices in HE institutions.

References

Crompton, H. (2013). A historical overview of mobile learning: Toward learner-centered education. In Z. L. Berge & L. Y. Muilenburg (Eds.), *Handbook of mobile learning* (pp. 3-14). Florence, KY: Routledge.

Crompton, H. (2017a). *2017 ISTE Standards for educators: A guide for teachers and other professionals*. Arlington, VA: International Society for Technology in Education.

Crompton, H. (2017b). Moving toward a mobile learning landscape: Presenting a mlearning integration framework. *Interactive Technology and Smart Education, 18*(2). doi: 10.1108/ITSE-02-2017-0018

NZMoE. (2012). *New Zealand curriculum update 26—future-oriented learning and teaching*. Wellington, New Zealand: Learning Media.

Oakley, A. (2012). Foreword. In D. Gough, S. Oliver, & J. Thomas (Eds.), *An introduction to systematic reviews* (pp. vii–x). London: SAGE Publications.

Sandelowski, M., Voils, C. J., Leeman, J., & Crandlee, J. L. (2011). Mapping the mixed methods—Mixed research synthesis terrain. *Journal of Mixed Methods Research, 6*(4), 317–331.

Statista. (2016). Share of mobile device owners worldwide from 2011 to 2016, by number of devices owned. Retrieved from https://www.statista.com/statistics/245501/multiple-mobile-device-ownership-worldwide/

Traxler, J. (2013). Mobile learning: Shaping the frontiers of learning technologies in global context. In R. Huang, Kinshuk, & J. Spector (Eds.), *Reshaping learning* (pp. 237–251). Berlin, Germany: Springer-Verlag.

Wright, G. B. (2011). Student-centered learning in higher education. *International Journal of Teaching and Learning in Higher Education, 23*(1), 92–97.

1

THE IPAD SIX YEARS ON

Progress and Problems for Enhancing Mobile Learning with Special Reference to Fieldwork Education

W. Brian Whalley, Alice L. Mauchline, Derek France, Julian Park and Katharine Welsh

Introduction

Fieldwork is an important part of research and educational activities in several disciplines. Geography, geology and bio- and ecological sciences are perhaps the most obvious, but this list can be extended to civil engineering and surveying, archaeology and history. Simply taking students out of the classroom can make their learning more engaging, as can being immersed in the environment that they are learning about. However, active learning is more than just being outdoors. Active learning includes any instructional method that engages students in the learning process by doing meaningful learning activities, especially thinking, questioning and problem-solving in the field (France, Whalley, & Mauchline, 2013). Examples of active learning approaches include problem-based learning, enquiry-based learning, types of peer-to-peer instruction and group/collaborative learning. These are in addition to traditional fieldwork attributes of observing, note-taking, data recording and interpretation.

The advent of the Personal Digital Assistant led to the smartphone, in particular Apple's iPhone, from 2007. A new dimension was added to the hardware with the rapid development of a wide variety of apps, some of which could be used for fieldwork enhancement, especially in higher education (HE). It was not until the arrival of Apple's iPad in April 2010 that screen sizes became suitable for reading and annotating PDFs as well as other visual academic tasks.

To enable data and information to be recorded in the field, the paper notebook has traditionally been a central part of the fieldwork equipment, with waterproof or water-resistant paper, for note-taking and sketching. Users of paper notebooks might also need various other devices for field-orientated tasks: camera, hand lens, compass and clinometer, light meter and/or sound meter. For the most part, the iPad can incorporate all these functions. Traditional, paper-based methods can be replaced or supplemented by an iPhone or iPad. Note-taking in the field can be performed or enhanced digitally, photographically or with audio recording. Images of paper notes can be used to back up fieldwork and added to a digital notebook. Additionally, the digital facilities in smartphones and tablets can be used to enhance the accessibility of such notes.

The use of computers in fieldwork is of long standing. Gardiner and Unwin (1986) extoled the virtues of using computers for fieldwork data analysis and report writing. At that stage, however, personal computers were large and, even if they could be used in the field, they were expensive and had a short battery life. Student use of computers tended to be in the laboratory or classroom after the day's fieldwork. Even the advent of inexpensive 'netbooks' from about 2007 did not solve many of the difficulties of 'field computing' until the arrival of the iPad and the widespread uptake of apps. The iPad is essentially an upgrade of Alan Kay's 'Dynabook' of 1972 but, with the use of owner-selected apps, it has some of the characteristics of Neal Stephenson's 'Young Lady's Illustrated Primer' (Stephenson, 1995) in that it grows with the owner/user.

The Enhancing Fieldwork Learning Project

The (UK) Higher Education Authority-funded project Enhancing Fieldwork Learning (EFL) started in 2010 with the aim of using mobile devices to help enhance fieldwork learning. This chapter reports some of our investigations, particularly in the areas of biology, geography, earth and environmental sciences (BioGEES) fieldwork. We have used iPads and, to a lesser extent, iPhones as a means of extending student participation in fieldwork, primarily because they were the market leader at the start of the project. We have continued with them in several forms because of the wide number of apps available. Other types of smart devices could equally be used appropriately in the situations we describe. The EFL project has been concerned with developing the use of the iPad from the original function of a media-consuming device to become part of a personal learning environment (PLE) for students. By using an iPad, together with appropriate apps and affordances in pedagogically sound ways, students can benefit from these devices to promote their own active education (McHaney, 2011).

In order to investigate the potential of tablet computers to enhance field teaching and learning we initially purchased six (original) iPads in 2010, but this has since been supplemented by several iPad Minis (versions 2 and 4), increasingly as departmental stock. We loaned these to geography and bioscience fieldwork practitioners for their separate field trips (both one-day and residential) to explore how tutors and students could use iPads to develop teaching activities and to gauge how students responded to the use of mobile technologies. Most BioGEES student fieldwork is done in small groups. We examined how this class set of 12 iPads was used to support group work (usually four people per group) in the field for day classes as well as one or two week-long residential trips in the UK at several locations and abroad (New York, Iceland). Students were responsible for taking the iPads to the venue from their home departments.

This progressive-use approach allowed us to explore the breadth of pedagogic possibilities provided by iPads and the use of a selection of education-related apps that we wanted the students to use for particular exercises. Initially, this was practitioner-led and the learning tasks were set prior to the trip. The students were not always able to spend much time familiarizing themselves with the devices before the fieldwork but all students had a chance to use them to support their learning in the field. The limited number of iPads did not present a problem because of the group work. Some apps were recommended to the students for particular tasks, but the students were also at liberty to explore and use other apps as required. This also allowed the students to explore the capabilities of iPads and appropriate them for their learning.

Students' Learning Spaces

Students work in a range of educational spaces (Savin-Baden, 2008) and fieldwork provides one such 'learning space'. The notion of 'person plus computer' means that we can take our cognitive learning spaces around with us (Gärdenfors, 2004). By extension, 'fieldwork' no longer means in remote locations but *anywhere* users are located: home, hospital, lecture theatre or laboratory. A 'Personal Learning Environment' (PLE) can be defined by an individual person, but can be greatly enhanced by the possession of a device such as an iPad (or Stephenson's 'Illustrated Primer') and supplemented by 'the cloud' and Wi-Fi. Not only can digital information be stored, but inter-connected repositories can be searched and information retrieved and (re-)transmitted in a manner that is far more cost-effective than in pre-iPhone days. For example, information literacy skills can be developed by students searching electronic databases for additional information to incorporate into reports. The EFL project capitalizes on these attributes of information storage, connectedness and sharing data and analysis in any location.

When a student joins a university or college, an institutional app may provide information about the local milieu, including maps, social scene, lecture theatre locations and perhaps lecture times. The student can use their PLE in this everyday sense and add their academic work to this mobile environment. Academic work might include suggested reading, notes on the reading, bibliographic information and work to be submitted to tutors. The iPad alone is not the PLE but part of the knowledge network, and is probably more useful than the local Virtual Learning Environment (VLE) because it is mobile and can be personalized. Information sharing is of course already highly developed by students at an informal level via social media. Moreover, iPads are highly portable and may well evolve with their owners such that social and academic lives can, with care, be part of any individual's overall education. We have developed this concept further where we refer to the iPad and apps as a *vade mecum* (Whalley, France, Mauchline, Welsh, & Park, 2016). We suggest that these ideas concerned with mobile technologies need to be further developed in all areas of HE.

Early Adopting Practitioners: Teaching Staff

We have examined both student and tutor/lecturing staff ('faculty' in the US) by questionnaires and focus groups on fieldwork events. At an early stage we asked fieldwork practitioners in the Bio-GEES disciplines about their use of mobile technologies to support learning in the field. Participants were surveyed by questionnaire in 2011 to help understand their practical and pedagogic motivations. The majority of the 89 respondents indicated that data processing (including data capture, analysis and sharing) and information and communications technology (ICT) skills development were their primary motivations (Welsh, Mauchline, Park, Whalley, & France, 2013). The main barriers to adoption were the cost and reliability of the equipment as well as concerns over staff and student capabilities. We also noticed (Welsh, France, Whalley, & Park, 2012) that students were worried about damaging their own or an institution-provided smartphone, yet rarely voiced the same concerns about very expensive specialist equipment used in the field such as a differential GPS.

The main fieldwork learning tasks teaching staff devised involved the use of iPads to: take photographs and videos, enable web browsing, enter raw data, collect GPS-linked data and geo-reference photographs (Welsh et al., 2012). Other activities included the use of Twitter for communication and the creation of short reflective videos. However, we concurred with a JISC report (Linsey et al., 2011) that students, 'do not use their personal technologies for learning unprompted'. However, we also found 12 per cent of our practitioner questionnaire responses about emerging technology (in 2010) indicated that devices will become interconnected in the field; thus: 'People expect to be able to work, learn and study whenever and wherever they want to' (Johnson, Adams, & Cummins, 2012). Our showcase events and publication (France et al., 2015) show how this may be achieved. The HEA funding allowed us to put on annual two-day events.

Initial Student Responses

At the beginning of the project, few students had used an iPad. In our mixed-methodology approach, 173 undergraduates on six fieldtrips responded to a questionnaire (Welsh et al., 2015) used to gain some idea of how students used iPads during fieldwork and the perceived benefits and drawbacks of their use. The iPad is designed to be intuitive to use and we found that students were able to adapt to working with these devices very quickly. The students reported that the multi-functionality of the iPad and its portability were the main benefits. There were some misgivings by students at first, that they might break or damage their loaned iPads, as well as some issues relating to internet connectivity availability in the field. To obviate damage to iPads we purchased protective cases for student use. This increases their bulk and weight a little and can be a nuisance when some accessories, such as supplementary lenses, need to be used. In fact, over the project we had no problems with damage and the cases ensured that one iPad survived being submersed in a geothermal pool on a trip to Iceland!

In our survey (Welsh et al., 2015) the students largely focused on the technology aspect of the device with 26 per cent recognizing that the iPad was convenient as an all-in-one or multi-tool device. They did not, however, recognize pedagogic benefits such as increased engagement and the iPads' use as a reflective tool. This was a similar finding to that of Woodcock et al. (2012), which suggested that students rely on guidance from their tutors in the use of mobile devices.

Some Examples of iPad-Supported Tasks in Fieldwork Learning

Although we found students took to using iPads with enthusiasm in their fieldwork, they frequently had to be told what facilities were available as affordances of the iPad and the apps that would be useful. We have listed many of these in France et al. (2015). Some other student experiences in using iPads are further discussed in Whalley et al. (2015). Although smartphones provide pedagogic affordances for some mobile devices, the larger screen offered by tablets provides much greater usability than small smartphones as it enhances the 'device aspect' of Koole's (2009) FRAME model.

That an iPad can act as a device for note-taking and photography, still and video, is only one aspect of their capabilities. Apps are available for making it easier to accomplish quite complex tasks, such as editing video or still photographs and audio dubbing. For some applications an instructor has to know the capabilities of the iPad as well as the apps available before going into the field. For example, in geological sciences it is often necessary to take angular measurements and direction of rock strata (dip and strike). The traditional method is to use a separate clinometer and compass, take many measurements in the field and then undertake the data reduction and plot it back at base. Only there was it possible to look at field sampling and the data's statistical significance. Now it is possible to take these measurements with an iPad in the field. Field measurements can be plotted at the site along with notes and a geo-referenced (GPS) location (see Figure 1.1). Not only does this give immediacy, it allows a more involved and active engagement in any field mapping programme. It also saves money on expensive hardware.

In general then, instructors need to know what can be done with iPads and apps and also to demonstrate these to students and for students to practice data gathering *before* going into the field. This applies to many apps, both general (such as word processors and spreadsheets) and for specific needs such as metadata editing, image processing and sketching aids. Skitch is one such app that can be used

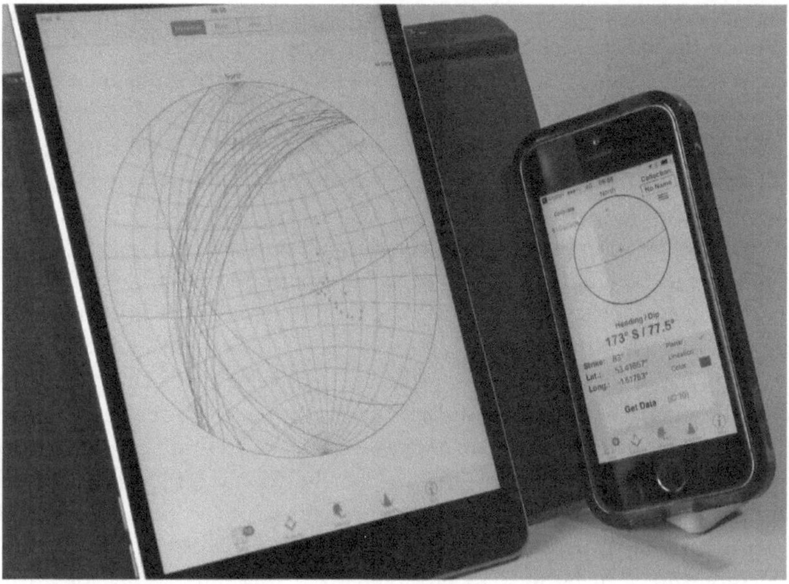

FIGURE 1.1 An app (Lambert) that allows the iPhone (as here) to be used as a clinometer and compass. In the field this is used to measure geological structures but here allows measurement to be practiced by measuring the slope of a book cover. This shows the principles of the measurement technique to be practiced before students venture into the field. The inset shows a set of collected data automatically plotted by the app. Both images were taken on an iPad Mini 4. W.B. Whalley 2016, CC BY-SA.

for drawing but also for annotating images and for practicing observation and supporting field sketching. Other case studies of using mobile technologies in the field can be found in France et al. (2013), and König and Bernsen (2014) provide an overview of iPad use from history education that includes student field trips.

Computer manufacturers have paid, rightly, attention to the 'accessibility' of their devices. The iPad can have different keyboards, both physical and virtual, to aid data entry and writing in the field. Cold wet fingers in the field can be left in gloves if audio notes are taken with the aid of a suitable app and a lapel ('lavalier') microphone. This technique for recording notes could be useful for students with certain impairments, as could the use of virtual assistants (Siri), whether in the field or not. Students with specific learning difficulties can use iPads to assist with a variety of tasks (Nuttall and Nuttall, 2013).

Crowdsourcing of Data in the Field and the Concept of Citizen Science

Citizen Science, 'scientific work undertaken by members of the general public, often in collaboration with or under the direction of professional scientists and scientific institutions' (Oxford English Dictionary), is becoming increasingly popular for fieldwork. Members of the public as well as students can use their tablets or smartphones to identify and record a wide variety of plants and animals. Some aids are on-board apps, but a website example is iSpot hosted by the Open University via the OPAL (Open Air Laboratories) network. The GPS facility on mobile devices can identify locations (geo-referencing) and thus be used to map occurrences, densities, etc. EpiCollect and Open Data Kit are two examples of open source apps that can be customized on both iOS and Android systems to enable field data recording. Not only does this activity provide training for field biologists and ecologists, it creates a database for future research. For example, the spread of ash dieback fungus can be tracked via the Living Ash project to produce data of national importance. Crowdsourcing and sharing of data can lead to other novel approaches to shared student use of iPads in the field. Examples are an institutional 'Twitterscavenge' to locate resources and by using Facebook to record road-kill ('Project Splatter'). Internationally, Zooniverse, 'people powered research', shows that anyone can be a researcher and the project links research and education in a wide variety of projects, many of them field based.

Questionnaires can be used to provide 'citizen data', specifically for urban field surveys in sociology, marketing and human geography. Students from the University of Chester have used iPads in this manner on field courses in New York. For locations away from the university's home country, it is usually advisable to purchase SIM cards locally to avoid high roaming charges. The advantages of quick data recording also apply to conflating the data and its analysis. Such skills in data management can be seen in a wide variety of tasks for which iPads can be used. For example, data validity and sampling strategy can be examined in the field rather than waiting until the home laboratory is reached. Some apps are designed specifically for data analysis. As well the 'traditional' use of spreadsheets, some apps can allow data to be plotted directly or by way of imported data as csv files. Such apps may be as important in the laboratory as in the field and go some way to support linkages between students, devices—such as iPads—and data collection as part of a personal learning environment.

Bring Your Own Device

The Bring Your Own Device (BYOD) movement has its origins in the constructionist ideas of the late Seymour Papert and Alan Kay and the One Laptop per Child (OLPC) initiative. Although the cost per item was used to criticize OLPC, the same might be said of the iPad. Cheaper tablet versions do now exist and the smartphone revolution has already arrived in the economies of Africa (Goldstuck, 2012). Economics as well as ease of use will probably further drive the move towards the acceptance of iPads and tablet/smartphone technologies and replace the classroom computer. As indicated previously, we have found that there is still some reluctance from many academics to use mobile technologies in their everyday teaching with students. There may be several reasons for this. Perhaps one is that if it is not a laptop then it cannot be a serious contender for student use, despite the entreaties of JISC (2015) to encourage mobile learning.

Some schools in the compulsory education sector have banned the use of smartphones, presumably because students are distracted by checking their social media. Conversely, some schools encourage the use of iPads and even produce guides for their use and suggest the apps to be installed. This is a move towards enlightened BYOD that we have yet to see in most HE institutions in the UK as they generally lack a BYOD policy. Some individuals have developed use of iPads as a means of transforming education, but the use of these devices needs to be encompassed within good design principles (Beetham, 2013; Kukulska-Hulme & Traxler, 2013) within HE institutions.

Personalizing Learning Environments

All of the features of iPad+apps and the affordances and abilities mentioned above allow a device to become much more than a field notebook. An iPad can act as a calendar, list maker, alarm clock and reminder tool; users can tailor their own devices to monitor their daily activities. This of course is what many people do who are not students in a formal sense: they play games, use email, log in to social media of many types on their smartphones and use instant payment technologies. We are now in Adam Osborne's vision: 'The future lies in designing and selling computers that people don't realize are computers at all'. iPads and other smart devices have become extensions of our minds and capabilities as indicated by Skyrme (1999) and can include aspects of Karl Popper's 'Three Worlds': the worlds of physical objects and events, mental objects and events and objective knowledge (Popper, 1999). Not the least of these is World 3: products of the human mind. As such, the iPad provides a personal and personalized link between individuals and local groups (students, work colleagues, family, clubs) as appropriate. We see this trend continuing and supporting mobile learning to counter some of the trends identified by Kukulska-Hulme (2012).

Institutional Advances since 2010

Since 2010, many HE institutions in the UK have started to support the integration of mobile learning into their educational programmes (JISC, 2015). Such support includes staff training, the development of staff-support mechanisms and the expansion of Wi-Fi coverage on campuses. Various possibilities for implementing iPad use and for supporting staff and students are discussed in the Proceedings of the 1st International Conference on the Use of iPads in Higher Education (Souleles & Pillar, 2015).

As well as promoting and investigating iPad use in fieldwork at our own institutions, we have run several workshops and six annual two-day 'showcase events' where academic practitioners have demonstrated their innovative pedagogies that utilize specific apps in fieldwork education. These events have been particularly useful in spreading the word about using iPads and exploring pedagogies for active fieldwork. Students too have been shown the value of the iPad to revolutionize their own learning and have sometimes developed ways to use their devices beyond learning tasks set by their tutors. We found that students, once exposed to mobile technologies, were able to link their use of specific apps to the graduate attributes that are important for employability (Hill, Walkington, & France, 2016). Institutional support and the greater availability of mobile devices together with the innovative ideas of practitioners has led to a drive towards novel pedagogies, especially those involving 'active learning' (Kukulska-Hulme & Traxler, 2013). This activity can be deployed in the lecture theatre as much as in fieldwork (Whalley, 2016).

Conclusions

Our, continuing, project has shown us the value for students of using mobile technologies in their out-of-classroom learning. This enhanced learning is also designed to be 'active' for students, and in the way that HE is situated as being 'research-led'. As with Citizen Science, anyone can participate in scientific investigation, students are part of this and assistance here is not 'what you know' as indicated on an informal level by quiz shows such as 'Mastermind', but, more formally, by the HE examination system. Rather, we take the approach of 'Research 3.0: driving the knowledge economy' of Maidment-Otlet and Redfearn (2010) in promoting mobile technologies, especially in the PLE. The iPad can be used effectively to enhance such education, not only in fieldwork but throughout education. Such breadth allows teaching practitioners and their students to follow the paths suggested in Laurillard (2013), Thomas and Seely Brown (2011) and Beetham and Sharpe (2013). As mobile smartphone and tablet technologies have developed rapidly over the last six years we believe there is no reason to think that the pace of development will slow. This applies to computing power, screen resolution and the inclusion of haptic devices and links to 'virtual' technologies as well as the apps that will be developed as a consequence. Student-centred learning, associated with the development of a PLE, should be used in most forms of education.

Acknowledgements

We acknowledge the initial funding of the UK Higher Education Academy and subsequently by the British Ecological Society for our Enhancing of Fieldwork Learning project. We also thank the contributors to our Enhancing of Fieldwork Learning showcase events.

References

Beetham, H. (2013). *Designing for learning in an uncertain future: Rethinking pedagogy for a digital age* (2nd ed.) (pp. 258–281). London: Routledge.

Beetham, H., & Sharpe, R. (2013). *Rethinking pedagogy for a digital age: Designing for 21st century learning* (2nd ed.). New York and London: Routledge.

France, D., Whalley, W. B., & Mauchline, A. L. (2013). Using mobile devices to enhance undergraduate field research. *CUR Quarterly, 34*(2), 38–42.

France, D., Whalley, W. B., Mauchline, A. L., Powell, V., Welsh, K. E., Lerczak, A., Park, J., & Bednarz, R. (2015). *Enhancing fieldwork learning using mobile technologies*. New York and London: Springer.

Gärdenfors, P. (2004). *Conceptual spaces: The geometry of thought*. Cambridge, MA and London: MIT Press.

Gardiner, V., & Unwin, D. J. (1986). Computers and the field class. *Journal of Geography in Higher Education, 10*(2), 169–179.

Goldstuck, A. (2012). Internet matters: The quiet engine of the South African economy. *World Wide Worx*, 38–50. Retrieved from http://hsf.org.za/resource-centre/focus/focus-66/AGoldstuck.pdf

Hill, J., Walkington, H., & France, D. (2016). Graduate attributes: Implications for higher education practice and policy: Introduction. *Journal of Geography in Higher Education, 40*(2), 155–163.

JISC. (2015). Mobile technology and m-learning. Retrieved from https://www.jisc.ac.uk/guides/using-assistive-and-accessible-technology-in-teaching-and-learning/mobile-technology-and-m-learning

Johnson, L., Adams, S., & Cummins, M. (2012). *The NMC Horizon report: 2012 higher education edition*. Austin, TX: The New Media Consortium.

König, A., & Bernsen, D. (2014). Mobile learning in history education. *Journal of Educational Media, Memory, and Society, 6*(1), 107–123.

Koole, M. L. (2009). A model for framing mobile learning. In M. Ally (Ed.), *Mobile Learning. Transforming the delivery of education and training* (pp. 25–47). Edmonton, Canada: AU Press.

Kukulska-Hulme, A. (2012). Smart devices or people? A mobile learning quandary. *International Journal of Learning, 4*(3–4), 73–77.

Kukulska-Hulme, A., & Traxler, J. (2013). Design principles for mobile learning. In H. Beetham & R. Sharpe (Eds.), *Rethinking pedagogy for a digital age: Designing for 21st century learning* (pp. 244–257). London: Routledge.

Laurillard, D. (2013). *Teaching as a design science: Building pedagogical patterns for learning and technology*. London: Routledge.

Linsey, T., Hall, R., Ooms, A., Taylor, P., Andrew, M., O'Brien, J., & Downward, S. (2011). Transforming curriculum delivery through technology: Stories of challenge, benefit and change. *JISC Final Report*. Retrieved from http://webarchive.nationalarchives.gov.uk/20140702233839/http://www.jisc.ac.uk/whatwedo/programmes/elearning/curriculumdelivery/curriculumdeliveryguide.aspx

McHaney, R. (2011). *The new digital shoreline*. Sterling, VA: Stylus.

Maidment-Otlet, D., & Redfearn, J. (2010). A research revolution: The impact of digital technologies. *Ariadne* (62). Retrieved from http://www.ariadne.ac.uk/issue62/maidmentOtlet-redfearn/

Nuttall, J., & Nuttall, L. (2013). *Dyslexia and the iPad*. CreateSpace Independent Publishing Platform.

Popper, K. (1999). *All life is problem solving*. London: Routledge.

Savin-Baden, M. (2008). *Learning spaces: Creating opportunities for knowledge creation in academic life*. Buckingham/Maidenhead: Open University Press, McGraw Hill.

Skyrme, D. (1999). *Knowledge networking: Creating the collaborative enterprise*. Oxford: Butterworth Heinemann.

Souleles, N., & Pillar, C. (Eds.). (2015). *iPads in Higher Education: Proceedings of the 1st International Conference on the Use of iPads in Higher Education (ihe2014)*: Cambridge Scholars Publishing.

Stephenson, N. (1995). *The Diamond Age*. New York: Bantam.

Thomas, D., & Seely Brown, J. (2011). *A new culture of learning: Cultivating the imagination for a world of constant change*, CreateSpace Independent Publishing Platform.

Welsh, K. E., France, D., Whalley, W. B., & Park, J. R. (2012). Geotagging photographs in student fieldwork. *Journal of Geography in Higher Education, 36*(3), 469-480.

Welsh, K. E., Mauchline, A. L., Park, J. R., Whalley, W. B., & France, D. (2013). Enhancing fieldwork learning with technology: Practitioner's perspectives. *Journal of Geography in Higher Education, 37*(3), 399–415.

Welsh, K. E., Mauchline, A. L., Powell, V., France, D., Park, J. R., & Whalley, W. B. (2015). Student perceptions of iPads as mobile learning devices for fieldwork. *Journal of Geography in Higher Education, 39*(3), 450–469. doi:10.1080/03098265.2015.1066315

Whalley, W. B. (2016). *Lecture discussion tasks: Pedagogic and technological approaches to help break down barriers in lectures.* Paper presented at the 6th International Symposium for Engineering Education, 2016, The University of Sheffield, July 2016, UK, Sheffield.

Whalley, W. B., France, D., Mauchline, A. L., Welsh, K. E., & Park, J. (2016). Everyday student use of iPads: A vade mecum for students' active learning. In B. Baab, J. Bansavich, N. Souleles, F. Loizides, & A. Mavri (Eds.), *Proceedings of the 2nd International Conference on the use of iPads in Higher Education (ihe2016)*, San Francisco (pp. 43–61). Cambridge Scholars Publishing.

Whalley, W. B., France, D., Park, J., Mauchline, A. L., Welsh, K. E., & Powell, V. (2015). iPad use in fieldwork: Formal and informal use to enhance pedagogic practice in a Bring Your Own Technology world. In N. Souleles & C. Pillar (Eds.), *Proceedings 1st International Conference on iPads in Higher Education (ihe 2014)* (pp. 100–112). Cambridge: Cambridge Scholars.

Woodcock, B., Middleton, A., & Nortcliffe, A. (2012). Considering the smartphone learner: An investigation into student interest in the use of personal technology to enhance their learning. *Student Engagement and Experience Journal, 1*(1), 1–15.

2

MAINSTREAMING MOBILE LEARNING IN JOURNALISM EDUCATION

Thomas D. Cochrane, Helen A. Sissons, and Danielle L. Mulrennan

Introduction: A Response to the Changing Nature of Journalism

University education across disciplines is falling short of the demands of "a dynamic and changing workforce" (A. Herrington & Herrington, 2006, p. 2). This includes journalism education. University journalism courses developed at a time when newspapers dominated the news landscape and, while this has not been the case for more than a decade, journalism education has been slow to react, with many only recently addressing the multimedia nature of twenty-first century journalism (Du & Thornburg, 2011) or including social media in their curricula (Folkerts, 2014).

The divergence between journalism education and the perceived requirements of newsrooms has meant criticism of the academy has been a feature of its relationship with industry going back to the 1960s (Dickson & Brandon, 2000; Du & Thornburg, 2011; Franklin & Mensing, 2010). Media professionals have traditionally expressed dissatisfaction with the writing and reporting skills of new graduates (Du & Thornburg, 2011), and universities have claimed their role was broader than simply training news reporters (Folkerts, 2014). Now editors want journalism educators to add to the foundational writing and reporting skills a knowledge of new technologies as these technologies emerge (Du & Thornburg, 2011).

This request for graduates capable of working with new technologies reflects the reality of newsrooms in the twenty-first century. Research over the past five years has shown that mobile technologies relevant to the journalism and news sectors have had a significant impact on the philosophies and therefore the operations of news media organizations worldwide. These organizations are now fragmenting and redefining themselves, while their journalists' professional roles have migrated from that of being gatekeepers of news to becoming participants, increasingly as co-creators, in a wider news conversation (Singer et al., 2011).

However, the technologies themselves are continuing to evolve at a dizzying rate, with new developments occurring constantly. Therefore, our research concluded that it was crucial to enable students to develop expertise in the use of social and mobile tools in an authentic and self-sustaining manner. Hence, we adopted a student-determined learning approach (heutagogy) on student-owned mobile devices to enable learning to move outside the classroom and continue beyond university.

Theoretical Background: Curriculum Design for Mlearning

Our focus is upon going beyond the frivolous and substitutionary use of mobile devices to reimagining mobile learning around designing transformative learning experiences that leverage the unique affordances of mobile learning. We believe that curriculum design should be driven by graduate outcomes—for example, matching the profile of a modern professional journalist—and founded upon a pedagogical strategy that is informed by the scholarship of technology-enhanced learning (SOTEL) (Cochrane & Narayan, 2015; Cochrane, Narayan, & Antonczak, 2015; Haynes, 2016; Wickens, 2006). Simply adding new technologies to pre-existing pedagogical practice will not achieve this:

> If we want students to become smarter than a smartphone, we need to think harder about the pedagogies we are using to teach them. Technology can amplify great teaching but great technology cannot replace poor teaching.
>
> *OECD, 2015, p. 4*

To achieve our aim we need to go beyond comparative case studies that invariably lead to no significant difference between traditional pedagogies and the substitution of mobile learning. As critiqued by Rushby (2012), "This type of research is characterised by: Small sample sizes, User acceptance, Positive results, i.e., the learning outcomes were usually slightly better or no worse than with other forms of e-learning".

Traxler (2016) has gone further in his critique of the state of mobile learning, asking, "What killed the mobile learning dream?" According to Traxler, "We've ended up with mobile access to virtual learning environments that are being used as repositories".

Cook and Santos (2016), are more positive about the potential of mobile learning, when aligned with a learning design approach that leverages new pedagogies: "A key evolving pedagogical affordance of mobile devices is the ability to use social media and apps to enable new patterns of connected social, learning and work-based practices". Building upon the work of Bannan, Cook and Pachler (2015), Cook and Santos argue for the application of Design Based Research (DBR) as a guiding methodology for curriculum design for mobile learning, as DBR deals with the messiness of mobile learning research and practice. DBR follows a cycle of informed exploration, enactment, evaluation, and redesign, leading to the development of transferable design principles. DBR provides a rigorous methodology for treating curriculum design as a design science as called for by Laurillard (2012) that goes beyond the comparative case study approach typical of much mobile learning research (Rushby, 2012).

Critical Incidents in Redesigning the Journalism Curriculum

Throughout our journey of exploring the impact of mobile social media on the journalism curriculum and the development of our mlearning design framework (see the end of this section) there have been several significant catalysts and critical incidents, including:

- 2011: Establishment of a Journalism Community of Practice (COP)
- 2012: Learning and Teaching Development Fellowship and expansion of COP, presentation at the Ascilite2012 Conference, and a Faculty Award for Excellence in Research Outcomes
- 2013: Development of Mobile Journalism (MoJo) course, presented at the International Conference on Information and Communication Technologies in Education (ICICTE Conference)
- 2014: Collaboration with Law School, introduction of blogging and social media portfolios into the first year curriculum
- 2015: Curriculum redesign of three-year journalism programme begins based upon heutagogy and the Pyramid of Journalism Competence (Clark, 2014)
- 2016: Beginning of integration of MoJo across the curriculum, showcase via World Journalism Education conference
- 2017: Launch of the new curriculum

Establishing a Journalism Lecturer Community of Practice

Born out of weekly discussions between a journalism lecturer and an academic advisor in 2011, we formed a core group of interested journalism lecturers in 2012 to explore the impact of mobile social media on the journalism profession and the journalism curriculum (Cochrane, Sissons, & Mulrennan, 2012). We met weekly at a local café and, enabled by free Wi-Fi and our mobile devices, we created a sustained community of practice that supported initial efforts at implementing new pedagogical strategies within our teaching practice. These practices integrated the use of student-owned mobile devices. We named the COP the Enhancing Journalism Education (EJE) Team, and we created a Wordpress blog to document our thoughts and experiences. We established a Google Plus (G+)

Community as an online network hub to broker our experiences to the wider department and begin forming a global network of interested educators.

Modelling the Professional Use of Mobile Social Media

One of the first things we did was conduct a survey of the journalism students' use of social media and mobile technologies. The survey revealed students were aware of tools such as Twitter and Facebook, but had no models of the professional use of mobile social media. This indicated that there was a gap between theory and practice in education. Thus a goal of the EJE Team COP was to model the professional use of these tools to the students, as well as to our teaching peers. In this way we hoped to encourage the students to use the tools and our colleagues to join us in modeling their use, including Twitter, blogging, Storify, and other emerging forms of online news media (Cochrane, Mulrennan, Sissons, Pamatatau, & Barnes, 2013; Cochrane, Sissons, et al., 2013). We found the most authentic way of achieving this was to look at obvious areas within the curriculum where the tools could be applied. For example, blogging was used in three different ways:

1. Live-blogging was introduced into the classroom using the educational tool Todaysmeet.com (https://todaysmeet.com/). By creating a room and inviting others in, students could live-blog a classroom discussion. It was valuable for the lecturers as they could archive the students' comments and reflect on them as a form of feedback on how the lesson went. We also found that students who were reluctant to contribute to oral discussions in the classroom would feel more comfortable "posting" a comment or question on Todaysmeet.
2. Reflective practice: From day one, the students set up their own blogging sites. This was a private blog, shared with their lecturer, that enabled students to critically reflect on their learning path and the work they were producing across their degree programme. We modelled Wordpress (http://wordpress.com) as it was consistent with the format of the journalism department's news website at the time. However, students were invited to use any blogging tool of their choice and some used Blogspot.
3. Microblogging: The confinement of 140 characters teaches brevity and sharpens a journalism student's ability to percolate an idea down to a short sentence and present it in an engaging way. Twitter was a useful tool for this, and the students also used Twitter direct messaging to communicate with lecturers when submitting potential story ideas.

The phases of story creation were reframed to reflect social media functions:

- Researching and information gathering, interviewing, and using crowdsourcing to capture contextual information were identified as "capture".

- Constructing a story, filtering out unnecessary information and prioritizing facts and writing a news narrative were identified as "curation".
- The publication of stories was identified as "sharing".

Exploring New Pedagogies, in Particular Heutagogy

Brokered by the academic advisor as a technology steward for the EJE Team COP, the team identified the relevance of heutagogy (Blaschke, 2012) as a foundational pedagogical driver for supporting the types of graduate skills that new media journalism requires (Cochrane, Antonczak, Guinibert, & Mulrennan, 2014; Cochrane, Mulrennan, et al., 2013; Mulrennan, Sissons, Rive, McMeekin, & Cochrane, 2015). We argue that teaching methods such as pedagogical (teacher-determined) and even andragogical (self-directed) approaches are increasingly becoming outmoded, in favour of the heutagogical (student-determined) learning approach, which has a net-centric theory that provides an effective framework for the education of journalism students in the digital age (Blaschke, 2012). This foundational framework features several key learning design principles:

- Recognition of the emergent nature of learning and hence the need for a "living" curriculum that is flexible and open to change as the learner learns;
- Related to this is the involvement of the learner in this "living" curriculum as the key driver;
- Recognising that knowledge and skill acquisition, and learning are separate processes and need different approaches;
- Identification of learning activities/processes by the learner not just the teacher.
- Using action research and action learning as meta-methodologies in the learning experience;
- Involvement of the learner in the design of assessment, self-diagnosis and application of knowledge in real life contexts;
- Collaborative learning;
- Coaching for individual learning needs and application.

Hase & Kenyon, 2007, pp. 114–115

By adopting a heutagogical teaching approach of student-determined learning, students can take a proactive role in the acquisition of skills necessary to utilize mobile social media tools in the production of news stories, thereby removing the disconnect between the theory and practice (Cochrane, Mulrennan, et al., 2013).

Development of a Mobile Journalism Course

In 2013 the EJE Team had the opportunity to put their ideas into practice within the curriculum through the redesign of an online journalism course. The course was renamed and redesigned as the Mobile Journalism course and became a vehicle for the embodiment of the EJE Team practice within the journalism curriculum

(Cochrane, Mulrennan, et al., 2013). Positive feedback from students, colleagues, and international peers (through various conference presentations and published articles) led to the ideas underpinning the MoJo course becoming core concepts within the five-yearly curriculum review process, and leading curriculum redesign and integration of mobile social media across the entire journalism curriculum.

Curriculum Redesign and Integration of Mobile Learning

Throughout 2015 and early 2016 the EJE Team led the process of redesigning the journalism curriculum (Cochrane, Sissons, Mulrennan, & Rive, 2016) based upon our now clearly defined mobile learning design framework (Table 2.1). The EJE Team formed two curriculum redesign teams, an undergraduate and a post-graduate team, each meeting weekly to discuss the implications for new learning outcomes and assessment designs for the journalism course that integrated the professional use of mobile social media and heutagogy into the curriculum. Minutes of these meetings and all new curriculum documents were collated and shared across the department lecturers via a shared Google Docs folder. The use of mobile social media was integrated into the annual introductory student orientation week. A day-long lecturer professional development workshop was also facilitated by the EJE Team at the start of 2016. This workshop focused upon department lecturers establishing their own mobile social media profiles.

Caveats

While great strides have been made in reimaging and redesigning the journalism curriculum through the integration of mobile learning and heutagogy, it has not all been plain sailing. We have found the biggest impediment to curriculum redesign is the reluctance of colleagues beyond the core members of the EJE Team COP to engage and implement change. While we have tried to broker the practice of the COP beyond its borders, bringing peripheral members into active participation is a very slow, and often confrontational, process. While all now agree that change in the profession of journalism, including the impact on journalism practice of mobile social media, means a change in journalism education is inevitable, it is a challenge to move lecturers beyond their prior teaching practice approaches. We believe a change in practice is first predicated upon a change of conception of what it means to be a teacher/lecturer in the twenty-first century, and this conceptual change can be painful to long-held and comfortable practices.

Mlearning Design Framework

Based upon our journey of exploring the redesign of the journalism curriculum over the past five years as outlined in the previous section, we argue that an authentic mobile learning design framework will integrate the SOTEL within a DBR methodology that transforms academic structures and practice. Key elements of this framework that have been identified include seven transformations of culture and practice:

- Transform academic culture: We need to move towards sharing best practice via open collaborative scholarship within global networks, embedding SOTEL as a critical element of teaching practice (Cochrane, Narayan, Antonczak, & Burcio-Martin, 2016; Haynes, 2016; Weller, 2014; Wickens, 2006).
- Transform our teaching practice and professional development towards a global network of COPs thus modelling professional participation within a global community. Reified, for example, as a connectivist Massive Open Online Course or cMOOC (Cochrane, Narayan, Burcio-Martin, Lees, & Diesfeld, 2015).
- Transform graduate outcomes from a focus on measuring competence to building capability and capacity to problem solve within teams and novel situations (Hase & Kenyon, 2007).
- Transform curriculum design from an individual academic exercise towards a collaborative curriculum design process (Laurillard et al., 2013).
- Transform assessment to authentic and student-determined scenarios and projects (J. Herrington, Reeves, & Oliver, 2009; Narayan & Herrington, 2014). This is encapsulated in the concept of rhizomatic learning (Cormier, 2008), where teachers become designers of an ecology of resources and triggering events that facilitate student creativity.
- Transform learning from passive to active participation in networked collaborative learning communities (Cochrane & Antonczak, 2015c).
- Transform Information Technology (IT) infrastructures and relationships to enable new pedagogical practice, moving from IT's role as controller of network access and capability to enabler of new ways of working and BYOD (Cochrane & Munn, 2016; Salmon & Angood, 2013).

Our findings led us to founding our curriculum redesign process upon a mobile learning design framework illustrated in Table 2.1 that maps the crossover between educational design research, learning design, design-thinking, and the relationship with theory, practice, and mobile learning across each of the four stages of a DBR methodology (Cochrane & Antonczak, 2015a; Cochrane & Narayan, 2017).

We believe that this mlearning design framework is transferable to educational contexts beyond journalism education, and we explore transferable design principles in the following section.

TABLE 2.1 Mlearning Design Framework

Methodology: (Educational) Design-Based Research	*Stage 1*	*Stage 2*	*Stage 3*	*Stage 4*
4 Stages of Learning Design	Informed Exploration	Enactment	Evaluation: Local Impact	Evaluation: Broader Impact
Connecting Theory and Practice	Theory	Practice	Participant Feedback	Critical Reflection
Intersection with Mobile Learning	Mobile Social Media Framework: Informing Curriculum Redesign	Rhizomatic Learning: Developing an Ecology Of Resources (EOR) Designing Triggering Events	Participant Feedback	Peer Reviewed Feedback via SOTEL
Design Thinking	Observe and Define	Ideate and Prototype	Iterative Testing and Improvement	Wider Testing

Transferable Design Principles

In this section we reflect upon the implications of our experience for broader educational contexts, identifying transferable design principles that we have begun to implement in contexts such as: Communication Studies, Health, Paramedicine, Law, Graphic Design, Public Relations, Fashion Design, Product Design, Carpentry, Boat Building, Landscape Design, and Game Development (Cochrane & Narayan, 2017).

Developing a Mobile Learning Strategy

Begin with the graduate profile. What are the key skills and attributes required? Clearly define the core competencies required by graduates, for example the Journalism Pyramid of eCompetence (http://www.poynter.org/2014/the-pyramid-of-journalism-competence-what-journalists-need-to-know/251048/) or developing Capability based upon Heutagogy:

> Capability is a holistic attribute and concerns the capacity to use one's competence in novel situations rather than just the familiar, a justified level of self-efficacy for dealing with novel problems, having appropriate values, being able to work in teams, and knowing how to learn.
>
> *Hase & Kenyon, 2007, p. 113*

Think creatively about how the graduate outcomes can be achieved. Design mobile learning experiences that can effectively engage and challenge students and enhance the traditional classroom experience. Focus upon the unique affordances of mobile learning: such as flexibility, collaboration and active participation in a global professional community. Enable students to be more integrated with the learning process by adopting a heutagogical approach whereby the students are encouraged to determine aspects of their learning experience.

Lecturers must develop and model professional mobile and social media presence and profiles.

Design triggering events that will stimulate student discussion and collaboration rather than delivery of content (Cochrane & Antonczak, 2015b; Cochrane, Antonczak, & Guinibert, 2014).

Design an ecology of resources that supports the integration of mobile learning:

- Course hub (for example: Learning Management System (LMS) or Wordpress)
- Community discussion forum (for example: G+ Community)
- Communication and networking (for example: Twitter hashtag, Slack)
- Reflective learning journals (for example: Wordpress, Mahara)
- Develop professional portfolios and profiles (for example: LinkedIn, Wordpress)
- Link to global experts via live and archived webinars (for example: live via Skype, live and archived by YouTube Live, Blackboard (BB) Collaborate)

Design authentic assessment activities that align the integration of mobile learning with the key graduate outcomes.

Conclusions

In this chapter we have outlined the critical incidents that have enabled the integration of mobile learning within the journalism curriculum based upon the development of a mobile learning design framework. This framework has in large part developed out of the experiences of a journalism lecturer COP that has spanned several years. We have based this design framework upon a DBR methodology that integrates the scholarship of technology-enhanced learning, and maps the curriculum design process to a four-stage learning design process. While the resistance to change among colleagues is something that will always be a characteristic when innovation is introduced, we believe the framework provides a clear design guide for moving forward, and that it will be useful in curriculum contexts beyond journalism. In a post-internet world where mobile device ownership is ubiquitous, education needs new models for curriculum design that authentically integrate the use of mobile devices in learning alongside the demands of industries' best contemporary practices.

References

Bannan, B., Cook, J., & Pachler, N. (2015). Reconceptualizing design research in the age of mobile learning. *Interactive Learning Environments*, 1–16. doi:10.1080/10494820.2015.1018911

Blaschke, L. M. (2012). Heutagogy and lifelong learning: A review of heutagogical practice and self-determined learning. *The International Review of Research in Open and Distance Learning, 13*(1), 56–71.

Clark, R. P. (2014, 12 May). The Pyramid of Journalism Competence: what journalists need to know. Retrieved from http://www.poynter.org/2014/the-pyramid-of-journalism-competence-what-journalists-need-to-know/251048/

Cochrane, T., & Antonczak, L. (2015a). Connecting the theory and practice of mobile learning: a framework for creative pedagogies using mobile social media. *Media Education, 6*(2), 248–269.

Cochrane, T., & Antonczak, L. (2015b). Designing creative learning environments. *Interaction Design and Architecture(s) Journal - IxD&A*, N. *24*, 125–144.

Cochrane, T., & Antonczak, L. (2015c, 14–16 March). Developing Students' Professional Digital Identity. Paper presented at the 11th International Conference on Mobile Learning, Madeira, Portugal.

Cochrane, T., Antonczak, L., & Guinibert, M. (2014, 24–26 November). Designing Transformative Learning Environments. Paper presented at Rhetoric and Reality: Critical perspectives on educational technology, the 31st Ascilite Conference, Otago University, Dunedin.

Cochrane, T., Antonczak, L., Guinibert, M., & Mulrennan, D. (2014, 28 February–2 March). Developing a mobile social media framework for creative pedagogies. Paper presented at the 10th International Conference on Mobile Learning, Madrid, Spain.

Cochrane, T., Mulrennan, D., Sissons, H., Pamatatau, R., & Barnes, L. (2013, 4–6 July). Mobilizing journalism education. Paper presented at the International Conference on Information Communication Technologies in Education (ICICTE 2013), Crete, Greece.

Cochrane, T., & Munn, J. (2016, 27–30 June). EDR and design thinking: Enabling creative pedagogies. Paper presented at the EdMedia: World Conference on Educational Media and Technology 2016, Vancouver, Canada.

Cochrane, T., & Narayan, V. (2015, 29 November–2 December). Developing the scholarship of technology enhanced learning (SOTEL). Paper presented at Globally Connected, Digitally Enabled, the 32nd Ascilite Conference, Curtin University, Perth.

Cochrane, T., & Narayan, V. (2017). Design considerations for mobile learning. In C. Reigeluth, B. J. Beatty, & R. Myers (Eds.), *Instructional-Design Theories and Models* (Vol. 4, pp. 385–414). New York: Routledge.

Cochrane, T., Narayan, V., & Antonczak, L. (2015, 19 November). Mobile learning, altmetrics, and SOTEL. Paper presented at the 5th Mobile Creativity and Innovation Symposium, RMIT, Melbourne, Australia.

Cochrane, T., Narayan, V., Antonczak, L., & Burcio-Martin, V. (2016, 19–20 April). Modelling open practices in professional development: Creating a culture of open social scholarship. Paper presented at OER16: Open Culture, University of Edinburgh, UK.

Cochrane, T., Narayan, V., Burcio-Martin, V., Lees, A., & Diesfeld, K. (2015, 29 November–2 December). Designing an authentic professional development cMOOC. Paper presented at Globally Connected, Digitally Enabled, the 32nd Ascilite Conference, Curtin University, Perth.

Cochrane, T., Sissons, H., & Mulrennan, D. (2012). Journalism 2.0: Exploring the impact of mobile and social media on journalism education. In I. A. Sánchez & P. Isaias (Eds.), *Proceedings of the IADIS International Conference on Mobile Learning 2012* (pp. 165–172). Berlin, Germany: IADIS International Association for Development of the Information Society.

Cochrane, T., Sissons, H., Mulrennan, D., & Pamatatau, R. (2013). Journalism 2.0: Exploring the impact of mobile and social media on journalism education. *International Journal of Mobile and Blended Learning, 5*(2), 22–38. doi:10.4018/jmbl.2013040102

Cochrane, T., Sissons, H., Mulrennan, D., & Rive, V. (2016). Journalism 2.0: Collaborative curriculum redesign. In D. Parsons (Ed.), *Mobile and Blended Learning Innovations for Improved Learning Outcomes* (pp. 181–200). Hershey, PA, USA: IGI Global.

Cook, J., & Santos, P. (2016). Three phases of mobile learning state of the art and case of mobile help seeking tool for the health care sector. In D. Churchill, J. Lu, T. K. F. Chiu, & B. Fox (Eds.), *Mobile Learning Design* (pp. 315–333). Singapore: Springer.

Cormier, D. (2008). Rhizomatic education: Community as curriculum. *Innovate, 4*(5), np. Retrieved from http://davecormier.com/edblog/2008/2006/2003/rhizomatic-education-community-as-curriculum/

Dickson, T., & Brandon, W. (2000). The gap between educators and professional journalists. *Journalism and Mass Communication Educator, 55*(3), 50–68.

Du, Y. R., & Thornburg, R. (2011). What is needed vs. what is taught: A national survey of online journalism instructors. Paper presented at the Small Program Interest Group at the AEJMC Annual Convention.

Folkerts, J. (2014). History of journalism education. *Journalism & Communication Monographs, 16*(4), 227–299.

Franklin, B., & Mensing, D. (2010). *Journalism education, training and employment*. New York: Taylor & Francis.

Hase, S., & Kenyon, C. (2007). Heutagogy: a child of complexity theory. *Complicity: an International Journal of Complexity and Education, 4*(1), 111–118.

Haynes, D. (2016). Introducing SOTEL. *International Journal for the Scholarship of Technology Enhanced Learning, 1*(1), 1–2.

Herrington, A., & Herrington, J. (2006). What is an authentic learning environment? In A. Herrington & J. Herrington (Eds.), *Authentic learning environments in higher education* (pp. 1–13). Hershey, PA: Information Science Publishing.

Herrington, J., Reeves, T., & Oliver, R. (2009). *A guide to authentic e-learning*. London and New York: Routledge.

Laurillard, D. (2012). *Teaching as a design science: Building pedagogical patterns for learning and technology*. New York: Routledge.

Laurillard, D., Charlton, P., Craft, B., Dimakopoulos, D., Ljubojevic, D., Magoulas, G., Masterman, E., Pujadas, R., Whitely, E. A., & Whittlestone, K. (2013). A constructionist learning environment for teachers to model learning designs. *Journal of Computer Assisted Learning, 29*(1), 15–30. Retrieved from http://dx.doi.org/10.1111/j.1365-2729.2011.00458.x doi:10.1111/j.1365-2729.2011.00458.x

Mulrennan, D., Sissons, H., Rive, V., McMeekin, S., & Cochrane, T. (2015). MoJoMLaw: Mobile social media integration in journalism and law. In S. Frielick (Ed.), *LTDF 2014* (pp. 70–72). Auckland University of Technology: Centre for Learning and Teaching.

Narayan, V., & Herrington, J. (2014, 24–26 November). Towards a theoretical mobile heutagogy framework. Paper presented at Rhetoric and Reality: Critical Perspectives on Educational Technology, the 31st Ascilite Conference, Otago University, Dunedin.

OECD. (2015). *Students, computers and learning*. Paris: OECD Publishing.

Rushby, N. (2012). Editorial: An agenda for mobile learning. *British Journal of Educational Technology, 43*(3), 355–356. doi:10.1111/j.1467-8535.2012.01313.x

Salmon, G., & Angood, R. (2013). Sleeping with the enemy. *British Journal of Educational Technology, 44*(6), 916–925. doi:10.1111/bjet.12097

Singer, J. B., Hermida, A., Domingo, D., Heinonen, A., Paulussen, S., Quandt, T., Reich, Z., & Vujnovic, M. (2011). *Participatory journalism: guarding open gates at online newspapers*. Malden, MA: Wiley-Blackwell.

Traxler, J. (2016). What killed the mobile learning dream? *Jisc Inform*. Retrieved from https://www.jisc.ac.uk/news/what-killed-the-mobile-learning-dream-26-feb-2016

Weller, M. (2014). *The battle for open: How openness won and why it doesn't feel like victory*. London: Ubiquity Press.

Wickens, R. (2006). SoTEL: Toward a scholarship of technology enhanced learning. *Canadian Journal of University Continuing Education, 32*(2), 21–41.

3

CREATING TIME AND RESPONSIVE DIMENSIONS IN SCIENCE WITH MOBILE TECHNOLOGY

Martin P. Khechara and Sara C. Smith

Introduction

Mobile learning (mlearning) is now used extensively in higher education (HE) (El-Hussain & Cronje, 2010). The use of this technology, most commonly represented by smartphones (Ofcom, 2015), allows approaches such as the flipped classroom or 'flipping' to be facilitated (Bishop & Verleger, 2013). Content is recorded and made available online before class through a mobile device (Bergman & Sams, 2012), leaving face-to-face sessions free for other activities that support learning. The use of the flipped approach has been shown to have a range of positive impacts on students (Smith, Brown, Purnell & Martin, 2015; Witton, 2016).

This chapter will review the findings of studies undertaken in a healthcare science programme that employed flipping. These investigations show that although the technology facilitates delivery it may not cause the changes seen in students *per se*. We hypothesise that it is what we term 'safe cognitive dimensions' that are important for students. These spaces within groups promote a deeper learning approach and perhaps cause fundamental changes at the level of the students' brain chemistry. We aim to show how the use of technology creates time and responsive dimensions in science education—a more receptive approach that reflects how students like to learn. Here we also begin to link the electronic to the physical, something that is yet to be addressed in mobile learning research.

Literature Review: Use of Mobile Technology for Learning in Medicine and the Healthcare Sciences

Mobile technologies have been used in a variety of disciplines within medicine and healthcare science. This section presents a selection of the current literature regarding how mobile technology is being used in these subject areas and provides context for the study presented later.

The medical professional requires a wide variety of knowledge to carry out their everyday tasks. Mobile devices are now increasingly being employed by medical students and physicians in the work place to access information relevant to their role. The technology can provide a form of dynamic scaffolding that is responsive to the learners, own developing knowledge and competence (Bullock, Dimond, Webb, Lovatt, Hardyman & Stacey, 2015).

Initially, in medical education mobile devices were used to deliver lecture content and textbooks (Woodill & Udell, 2011). Internet audio broadcasts (podcasts) were subsequently introduced as an important way to deliver content (Long & Edwards, 2010). With the increase in availability of connected technology there is an increase in the use of task-specific mobile applications, or 'apps', to support learning. Medical education apps are now prolific (Woodill & Udell, 2011). The types of apps range from educational programs such as the Visible Body (www.visiblebody.com) to resources such as Daily Rounds for doctors (Sebin, 2015). In their study to investigate how an app could support the practice of newly qualified doctors in the United Kingdom (UK), Bullock et al. (2015) established that physicians responded positively to the increased access to up-to-date textbooks afforded by the app. The important point is that the app was seen to supplement learning and surprisingly also increased the interaction of junior doctors with more senior medical staff. Although the app was providing content it was not a replacement for human interaction; it enhanced participation in the community of practice (Bullock et al., 2015).

The use of mobile technology for learning across other healthcare disciplines is similar due to the nature of the factual material to be accessed. However, the use of mobile technologies to facilitate the flipped classroom is one area that has been evaluated widely in this setting (Pierce & Fox, 2012).

The subject area of pharmacy encompasses training to prepare individuals for eventual practice as a registered pharmacist. As with other disciplines, students are required to apply their learning in a variety of contexts. This requires a deep approach to learning that cannot be achieved through simple didactic delivery; an active learning strategy is required (Schneider, Munro & Krishnan, 2014). To achieve this, Schneider et al. (2014) used a flipped classroom approach, which provided more time for students to engage in discussions and problem-solving exercises during their lesson. They found that although the different teaching style motivated students to engage, there was no subsequent increase in student attainment in summative assessments. The major benefit reported was from the increased interaction with peers, enabling peer learning to take place (Schneider et al., 2014).

In contrast, Pierce and Fox (2012) who also used flipping for pharmacy education to provide time and space in the classroom for assessing and developing student knowledge, reported an increase in student performance. Similar findings were identified by Missildine, Fountain, Summers and Gosselin (2013), who compared nurses' exam performance in flipped classrooms against their performance in traditional ones. The researchers identified higher exam scores as well as greater student satisfaction scores for the flipped approach. Others have evaluated the wider impact of flipping on student learning and professional development, identifying its role in promoting team working skills essential in the healthcare setting (Clark, 2014).

Biomedical science is a multidisciplinary laboratory-based role within the healthcare sciences. In the UK, biomedical scientists often work within the National Health Service (NHS) and support diagnosis of disease in samples taken from patients. This is a demanding subject area where students must relate theory to practice and apply rigorous scientific method in the laboratory. As such, the biomedical science classroom faces many of the challenges other medical sciences do, including how to create a deep approach to learning. To achieve this in their classroom, Ernst, Harrison and Griffin (2013) created a smartphone app called the Virtual Human Body (VHB). Students engaged with the VHB using a variety of different devices and acknowledged that the app helped them to prepare for their final assessments. They also appreciated the go anywhere, learn at your own pace, anytime nature of the programme, and the authors demonstrated that the use of the app enabled a deeper approach to learning in their students (Ernst et al., 2013).

Each of the studies presented here has shown that the use of mobile technology in medicine and the healthcare sciences can allow the formation not only of communities of practice but also of communities of learning. It supports the creation of space and time within the classroom for interactions to occur beyond that found in the traditional classroom and increases engagement with content. It allows the formation of relationships and gives students the opportunity to interact with tutors at a collaborative level. In summary, the use of mobile technology, and the different teaching methods it can afford, leads to a range of valuable changes in students, which are not just based upon attainment in exams.

The study outlined in the following section addresses how we have used mobile technologies, their impact upon our approaches to teaching, and their impact upon students' perceptions and attitudes to learning within these classrooms.

Methodologies, Methods and Approaches Taken in the Studies Reviewed

The data used here was taken from three different studies at both undergraduate and postgraduate levels on a biomedical science education pathway. The studies were all firmly grounded in a socially constructivist paradigm (Andrews, 2012) where learning is seen as a collaborative event. This reflects the nature of training in healthcare sciences as capability is developed by interaction with a community of practice and as such is a valid epistemological standpoint for the work undertaken. As the researchers involved in the studies are practitioner researchers and part of the academic team on the programmes where the studies took place, an action research methodology was also employed (Open University, 2005).

Study one was carried out at postgraduate level and involved the use of the flipped classroom only. Students in this study did not use technology in the classroom but did access the flipped content on mobile devices before the session (Smith, Brown, Purnell & Martin, 2015).

Study two built on the findings of study one and also took a flipped approach at postgraduate level. However, in this case students were also encouraged to use mobiles in the classroom to review progress using the app Socrative (www.socrative.com) and to use their devices for internet searches during group work. The Bring Your Own Device (BYOD) (Sharples et al., 2014) approach to mobile learning was positively encouraged in these sessions where appropriate. The functioning group ensured that all students had access to any content needed even if they did not have a device of their own (Smith & Khechara, 2016).

The third study was carried out at undergraduate level and used the flipped approach to deliver instructional content to aid in understanding of practical laboratory classes. Mobile devices specifically for use in the laboratory environment were supplied in the form of a tablet device (Smith, O'Gara & Khechara, 2016).

Each of the studies used lecture capture software such as Panopto (www.panopto.com) to record and distribute the flipped content. This software allows the academic to record high definition (HD) video and sound, including slide or screen capture, as well as feeds from other camera-enabled devices such as microscopes. This is all facilitated by the operator on a stand-alone device, such as a desktop or laptop computer with a web cam and suitable microphone. Recordings are then delivered to the viewer through an online player accessed via a web link.

In all cases students' perceptions were evaluated through the use of semi-structured focus groups. Discourse from these sessions was electronically recorded and transcribed verbatim. For the purposes of this review, transcripts from focus groups from the three studies undertaken have been recoded using thematic analysis (Braun & Clarke, 2006) to allow a qualitative comparison of the data in each of the three projects undertaken.

Findings and Reflections

Themes arising from transcribed discourse were identified and a number of different interrelated codes established. These represented aspects of how the activities and use of mobile devices changed students personally, how these modified attitudes to work or modified attitudes towards the subject area.

For the postgraduate students, the flipped approach was acknowledged as allowing the creation of extra time in the classroom. This is highlighted in the excerpt from the first focus group, where students acknowledged the benefit of having more time with their tutors: 'You only have 3–4 hours in a whole week [contact hours per module]. You can use that time to ask questions and do more practical application of what you have learnt rather than just listening to theory' (Focus group 1, PK).

These findings reiterate the work of others who have demonstrated that flipping provides more usable time in the classroom (Pierce & Fox, 2012). It is this perceived extra time that enables a range of approaches to be adopted and so allows for the variety of changes we identified in students. The provision of more 'class time' in the postgraduate modules provided students with time to think, reflect and, most importantly, to explore. Here we propose the term 'cognitive time machine' to express how the flipped classroom creates time and space for the active learning strategies to be employed and the development of higher order thought processes that ultimately lead to deep learning strategies in students.

The flipped classroom is underpinned by the initial provision of material to be accessed via mobile devices, in particular the students' own devices. This socio-culturally appropriate use of technology to facilitate teaching and learning directly increased student engagement with the subject area from the start as demonstrated by the comments of one student outlining their approach to accessing the material: 'If I am watching a video on my phone, I am more engaged, as it's more interesting to see images and figures on my phone. It's like a movie so I'm more engaged' (Focus group 1, HS).

The power of the method to promote engagement was seen to occur in spite of a student's initial perceptions of the subject area and clearly shows the effectiveness of reversed pedagogies. One student clearly articulated this when discussing their thoughts on the flipped classroom: 'This approach awakes your interest even if you are not interested in it [the subject]' (Focus group 2, CW).

Increased engagement by students resulted in a greater understanding of the subject content and in turn facilitated further engagement. The format and ability to interact with content to learn at their own pace was identified as an advantage: This excerpt was a typical response from students in all three studies: 'The best thing about it is that you can pause and rewind, so if you don't understand something you can go back and look at it again, where as in class you don't feel like you should interrupt' (Focus group 2, LJ).

As the students engage and their understanding increases they see the value of a different approach to lesson structure and learning tasks. Indeed, all students from the three studies preferred the flipped delivery and thought the changes bought about in the classroom were positive.

The effects outlined relate to changes in how students engage with information and the teaching approach itself. However, the effect of the cognitive time generated can also be seen as a change in the attitudes of students towards themselves or others. Time in class where a student-centric pedagogy prevailed and group work could take place, facilitated by an academic, allowed relationships to form and grow between peers and between the students and teachers. This is seen as a valuable part of the learning experience as extra direct contact time could be used to explore problems with tutors in greater depth and students felt they received more support. Students commented on how they valued the extra support provided in the flipped classrooms and their perception that this time was of more value to them: 'You feel that you are getting more help and support from staff, from the lecturers; you feel that they are able to give you the support that you need' (Focus group 2, LC).

Growth in relationships in the classroom through group-working activities helped to build the confidence of students and further reinforced their ability to form relationships. This made students more comfortable to interact with activities such as presentations. This relationship building was clearly identified by the students in each of the focus groups, though it was more evident in the two groups that used mobile devices for in-class activities; these activities appeared to support the relationship building. The impact is evident from the comment below:

> [Classroom activities] … With presentations in the activities of the flipped, you would normally be nervous, but now we have got to know our peers, it's a lot easier, because we have made a relationship with them through interactive learning, in that sense you feel more confident standing up and presenting.
>
> *Focus group 2, LJ*

The use of mobile devices in the laboratory was also seen to increase students' motivation to interact. Here the flipped approach was adopted differently by students. Although intended to provide pre-class instruction, the majority of students accessed the instructions and demonstrations of techniques provided during the class time as required. This meant they did not have to ask for or wait for extra support from tutors. This further reinforced the emotional changes that we had identified in the previous studies were enabled via a flipped approach. One student commented: '[Flipped laboratory] … I think it gives you the motivation to literally study, to put your all into it' (Focus group 3, PB).

This identifies the positive impact on student attitudes. There was an increased motivation to engage in the topic and the learning activities in the laboratory. Students were able to watch the demonstrations on the tablets on their own or as a group to support their activities and progress through the practical class. A positive effect on the students' level of confidence was seen and it also reinforced their ability to build relationships through this shared experience.

In summary, the changes in attitudes and emotional state of students bought about by the flipped classroom and the active learning approaches used generated a deeper level of learning. Engagement in the topic content and a greater level of understanding when combined with the relaxed nature of the group, better relationships and motivation may be the driver for this change in cognitive processing. However, we feel that a more fundamental force is at work that underpins everything and affords students the cognitive space in which to change. This is the formation of safe spaces or what we have termed 'safe cognitive dimensions'.

In educational terms a safe space has been defined as: 'An environment in which everyone feels comfortable expressing themselves and participating fully, without fear of attack, ridicule or denial of experience' (Arao & Clemens, 2013, pp. 135–151). Hockings (2011) refers to safe spaces as safe inclusive spaces. She suggests that the HE classroom is often an inhospitable place focusing on delivery rather than dialogue between those in the classroom. This is particularly the case when students are taught in large classrooms where there is little opportunity for relationships to grow. Students rarely speak in these spaces for fear of appearing stupid. The environment has a numbing effect on the mind and promotes non-engagement. However, safe cognitive dimensions for learning can be established by developing relationships with students (Hockings, 2011).

In the studies reviewed the active learning approaches used in the flipped classroom addressed some of the negative factors experienced in the traditional classroom setting and allowed the formation of relationships consequently generating a feeling of safety.

We posit here that a safe cognitive dimension is a mental construct of the students' minds and is related to the level of psychosocial stress experienced. Psychosocial stress results from being confronted with perceived social threats that we think we may not have the resources to deal with mentally (Scott, 2016). Prolonged exposure to this type of stress may result in poor performance, disinterest, dejection and memory disturbances (Danielsson, Heimerson, Lundburg, Perski, Stefansson and Åkerstedt, 2012). Indeed stress has been shown to impair memory retrieval and is acknowledged to inhibit memory formation and learning particularly in a region of the brain called the hippocampus where memories are processed and stored (Kuhlman, Piel & Wolf, 2005; Mohapel, Leanza, Kokaia & Lindvall, 2005). This in part occurs through the action of a hormone called cortisol that is produced in response to stress (Kalat, 2007; Wolf, 2009). These factors taken together begin to explain the basis of what we see in the classroom during flipped sessions. A reduction in psychosocial stress experienced by the students working in a group should lead to a reduction in stress hormone production and consequently allow better memory formation and recall, and allow better learning (Clark, Kirschner & Sweller, 2012).

Mobile Technology and the Flipped Classroom: Concluding Reflections

We have seen that mobile technology can be used in a variety of ways within medicine and the healthcare sciences to support learning. However, the introduction of mobile or another kinds of technology into the classroom is not a simple panacea that can 'fix' the learning process. Technology is just a tool that without reflection on teaching practice by academics using it may have a negative effect (Carter, Greenberg & Walker, 2016). The use of the flipped classroom shows this particularly as the delivery of recorded didactic content cannot simply be used as a replacement for good teaching practice. Active learning strategies must be employed. This takes time and effort on the part of the academic to implement and facilitate, and without this the effectiveness of the approach is lessened.

Theories of Time and the Psychoneurobiological Effects of the Flipped Classroom

Here we have tried to unite the technological, the pedagogical and the biological in an attempt to present a theory of how the flipped approach leads to improved learning. Our results suggest that the use of the flipped classroom when facilitated, as in our case, by mobile devices develops safe cognitive dimensions through the creation of usable time. The method represents a cognitive time machine, literally creating time where it was limited before, allowing time to think, build relationships, develop understanding and produce a more engaged student. This all reinforces a sense of safety that reduces stress to allow the deeper approach to learning seen in these sessions.

Further Work

The role of stress and its effect on learning is very controversial (Wolf, 2009). Changes in stress levels in students exposed to traditional or flipped environments could be assessed indirectly through measures such as the UWIST test (Matthews, Jones & Chamberlain, 1990) or directly through the measurement of student salivary cortisol levels. Further work is certainly warranted in this area to determine whether a definite link to the effects of psychosocial stress and learning in the flipped classroom can be made.

Final Conclusions

In conclusion, we have shown the flipped classroom supported by mobile technology has an impact on students. It affects how they learn and perhaps even modifies their brains to learn more effectively. As practitioner researchers we continue to change our practice through the use of these pedagogies, and as we move into the future of a more rigorous funding structure in UK HE we will continue to use them to create time and responsive dimensions in science.

References

Andrews, T. (2012). What is social constructionism? *The Grounded Theory Review, 11*(1), 39–46.

Arao, B. & Clemens, K. (2013). From safe space to brave spaces. In L. M. Landreman (Ed.), *Art of effective facilitation: Reflections from social justice educators* (1st ed.) (pp. 135–151). Stirling, VA: Stylus.

Bergman, J. & Sams, A. (2012). *Flip your classroom. Reach every student in every class every day.* Washington, DC: International Society for Technology in Education.

Bishop, J. L. & Verleger, M. A. (2013). The flipped classroom: A survey of the research. ASEE National Conference Proceedings, Atlanta, GA.

Braun, V. & Clarke, V. (2006). Using thematic analysis in psychology. *Qualitative Research in Psychology, 3*(2), 77–101.

Bullock, A., Dimond, R., Webb, K., Lovatt, J., Hardyman, W. & Stacey, M. (2015). How a mobile app supports the learning and practice of newly qualified doctors in the UK: an intervention study. *BMC Medical Education, 15*(71), 1–6.

Carter, S. P., Greenberg, K. & Walker, M. S. (2016). The impact of computer usage on academic performance: Evidence from a randomised trial at the United States Military Academy. *School Effectiveness and Inequality Initiative*, 1–30.

Clarke, K. (2014). Flipping out: A trend in radiologic science education. *Radiologic Technology, 85*(6), 685–687.

Clark, R. E., Kirschner, P. A. & Sweller, J. (2012). Putting students on the path to learning. The case for fully guided instruction. *American Educator*, 6–11.

Danielsson, M., Heimerson, I., Lundburg. U., Perski, A., Stefansson, C. & Åkerstedt, T. (2012). Psychosocial stress and health problems. Health in Sweden: The national public health report 2012. Chapter 6. *Scandinavian Journal of Public Health, 40*(9), 121–134.

El-Hussain, M. O. M. & Cronje, J. C. (2010). Defining mobile learning in the higher education landscape. *Educational Technology and Society, 13*(3), 12–21.

Ernst, H., Harrison, J., & Griffin, D. (2013). Anywhere, anytime, with any device: scenario-based mobile learning in biomedical sciences. *International Journal of Mobile Learning and Organisation*, 7(2), 99–111.

Hockings, C. (2011). Hearing voices, creating spaces: the craft of the 'artisan teacher' in a mass higher education system. *Critical Studies in Education, 52*(2), 191–205.

Kalat, J. W. (2007). Stress and health. In *Biological psychology* (pp. 376–377). Belmont, CA: Thompson Wadsworth.

Kuhlman, S., Piel, M. & Wolf, O. T. (2005). Impaired memory retrieval after psychosocial stress in healthy young men. *The Journal of Neuroscience, 25*(11), 2977–2977.

Long, S., & Edwards, P. (2010). Podcasting: Making waves in millennial education. *Journal for Nurses in Staff Development, 26*(3), 96–101.

Matthews, G., Jones, D. & Chamberlain, A. (1990). Refining the measurement of mood: The UWIST Mood Adjective Checklist. *British Journal of Psychology, 81*(1), 17–42.

Missildine, K., Fountain, R., Summers, L. & Gosselin, K. (2013). Flipping the classroom to improve student performance and satisfaction. *Journal of Nursing Education, 52*(10), 597–599.

Mohapel, P., Leanza, G., Kokaia, M. & Lindvall, O. (2005). Forebrain acetylcholine regulates adult hippocampal neurogenesis and learning. *Neurobiology of Aging, 26*(6), 939–946.

Ofcom. (2015). The communications market report. Retrieved from http://stakeholders.ofcom.org.uk/binaries/research/cmr/cmr15/CMR_UK_2015.pdf

Open University. (2005). Action research a guide for associate lecturers. Centre for Outcome Based Education. Retrieved from http://www.open.ac.uk/cobe/docs/AR-Guide-final.pdf

Pierce, R. & Fox, J. (2012). Instructional design and assessment. Vodcasts and active-learning exercises in a flipped classroom model of a renal pharmacology module. *American Journal of Pharmaceutical Education, 76*(10), 1–5.

Schneider, J., Munro, I. & Krishnan, S. (2014). Flipping the classroom for pharmacokinetics. *American Journal of Educational Research, 2*(12), 1225–1229.

Scott, E. (2016). What is psychosocial stress. Retrieved from https://www.verywell.com/what-is-psychosocial-stress-3145133

Sebin, D. (2015, May 4). Top 10 must have medical apps for medical students and doctors. [Weblog comment] Retrieved from https://www.dailyrounds.org/blog/top-10-must-have-medical-apps-for-medical-students-and-doctors-for-usmle-preparation/

Sharples, M., Adams, A., Ferguson, R., Gaved, M., McAndrew, P., Rienties, B., Weller, M. & Whitelock, D. (2014). *Innovating pedagogy 2014: Open university innovation report 3*. Milton Keynes: The Open University.

Smith, S. & Khechara, M. (2016). 'Technologizing' the postgraduate classroom. In *Edulearn16*. Barcelona: Edulearn.

Smith, S., O'Gara, E. & Khechara, M. (2016). Developing student capability in a biomedical science award – peer supported learning through video. In *Edulearn16*. Barcelona: Edulearn.

Smith, S., Brown, D., Purnell, E. & Martin, J. (2015). 'Flipping' the postgraduate classroom: supporting the student experience. In P. Layne & P. Lake (Eds.), *Global innovation of teaching and learning in higher education. Transgressing boundaries* (pp. 295–316). Series: Professional Learning and Development in Schools and Higher Education, Vol. *11*. Switzerland: Springer Publishing International.

Witton, G. (2016). The value of capture: Taking an alternative approach to using lecture capture technologies for increased impact on student learning and engagement. *British Journal of Educational Technology*, *48*(4), 1010–1019.

Wolf, O. T. (2009). Stress and memory in humans: Twelve years of progress? *Brain Research*, *1293*, 142–154.

Woodill, G. & Udell, C. (2011). Mobile learning in medicine and healthcare: professional education applications. Float mobile learning. Retrieved from http://info.floatlearning.com/hs-fs/hub/241955/file-285652672-pdf/docs/mobile-health/mobile-learning-medicine-healthcare-professional-education-applications.pdf

4

AUGMENTED REALITY AND MOBILE LEARNING

Theoretical Foundations and Implementation

Andrew T. Greenwood and Minjuan Wang

Introduction

Technology continues to change the way we interact with the world. Mobile devices connect and empower us in ways that would have been difficult to imagine a few decades ago. Compared to even a few years ago mobile devices have grown exponentially in capability. This integration of technology into every facet of our daily activity allows us to extend our human capabilities and develop knowledge in ways not previously possible. This phenomenon has profound implications for education and training. For example, augmented reality (AR) affords the ability to overlay images, text, video and audio onto existing structures, space or imagery in real time and in context with the learner's environment. The engaging nature of AR makes it an especially attractive tool for education. AR normally is used in real time and in context with environmental elements. By incorporating virtual and real world experiences AR has many potential applications in education and product development.

In this chapter we will discuss the use of AR in education, introduce a theoretical foundation and address the challenges and barriers associated with AR's large-scale implementation and wider institutional acceptance. In particular, we focus on the theoretical aspects of AR and mobile AR as a way to analyze its potential integration in educational settings. While there are many applications, we will briefly discuss current research and application of AR in education, and state the case for the inclusion of these technologies as an integral approach for education.

Augmented Reality: Why Now?

Attempts at bringing virtual reality (VR) to the mainstream began over thirty years ago, with the strongest wave in the mid-1990s. Unfortunately, user experience back then did not live up to the publicity; the headsets were expensive, unwieldy and caused headaches; the graphics making up the virtual worlds were unconvincing; and a dearth of developers meant minimal options for content. Today, a convergence of favorable conditions is pushing AR and VR at an exponential rate. The developments in the last decade for mobile devices have put the technology in the hands of many for AR, with expanding development for VR.

Technology

Computer Graphics—For VR use, graphics have improved a great deal and are now easier to create and render. They are so realistic that virtual worlds and objects are more convincing, resulting in a truer environment for learning skills. AR graphic interfaces smoothly interact with real world triggers or canvases, which supports far transfer of skills, which refers to students' ability to use their skills from a familiar context to an unfamiliar one.

Headsets and Wearable Devices—All are smaller, lighter and more portable. VR environments now are less likely to cause nausea. Current technology, as well as regular improvements, make the visuals more realistic. Audio now incorporates binaural audio, which can provide a more fully immersive audio experience where you can hear sounds from 'behind' you as well as from other directions. Haptic gloves and handles mean you can interact and 'feel' the environment in realistic ways.

Processing—Technology has previously not been able to process the amount of information necessary for a user to experience VR or AR in real time. As the user turns around and interacts in the virtual world, the visuals have to be able to keep up. In the past, no processors had enough power to handle all of the math and 3D imaging. Not only can computers process the necessary information, mobile devices can support current applications of AR and many VR functions as well.

Cost to Consumers

The cost of mobile devices continues to drop as new technology enters the marketplace. Each consumer will get more out of each purchase (more things to do with the headset, more games, more entertainment), so individuals are more likely to engage, own and learn.

Investment

Between Google's $500 million investment into Magic Leap, and Facebook creating a whole division for VR, AR/VR is entering the mainstream. It is not an *if* but a *when*—and the when is arguably now. Across the board, development is happening. It will be what students expect—it will be how they *want* to learn. It will be what they're used to.

Implementation

Adding technology to curriculum necessitates infrastructure, support and training for successful implementation. Computers, headsets, software and associated devices demand technology-savvy custodians in order to remain operational. There are numerous points where an instructor or learner may encounter technology glitches or experience gaps in knowledge, which will lead to abandonment of the tool. Technical support will be needed to maintain the software, hardware and devices. How the hardware is managed and implications for securing the device will need to be addressed as well. Security issues with content could also be a consideration. In any integration of new technology the educators need to know how to use the devices purposefully. Educators will need to be taught how to use the technology, and how to implement it into their existing curriculum.

Theoretical Foundations

When discussing Augmented Reality in education, it is important to understand how it fits into how we view our reality. Mixed reality can be applied to any technology in a fully virtual environment. Mixed reality allows the interactions that occur between real and virtual objects.

Figure 4.1 displays the reality–virtuality (RV) continuum as proposed by Milgram and Kishino (1994). Augmented Reality merges virtual objects into the real world, while augmented virtuality (AV) puts real objects into virtual worlds. AV is most evident in playing games through popular consuls like Microsoft Xbox using the Kinect Camera, or through the Nintendo Wii with its motion controllers. Users become the controller through their actions and gestures and have them portrayed onto a virtual character or avatar. It is through the coordination of perception, cognition and interaction that arises in AR and AV platforms that objects are turned into tools for cognition and learning (Zap & Code, 2016).

It is important to consider how technology used in daily life can be used in education and design. Looking at technology outside of an academic environment,

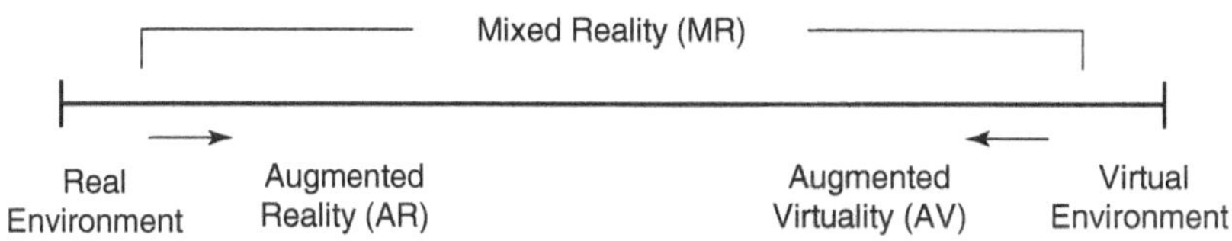

FIGURE 4.1 The Reality–Virtuality Continuum.

shifts are seen in learning styles happening in the students' informal, voluntary educational activities. When learners use their smartphones they infuse virtual resources as they move around in the real world. Using devices to text friends, sending pictures across distances, and accessing audio or video files are all frequent and common uses for these devices. In their learning processes, many of these socially created distributed communities parallel the activities of twenty-first century professionals in knowledge-based workplaces. Augmented Reality as a system can enhance a person's primary senses (visual, aural and tactile) with virtual or naturally invisible information made visible by digital means (Chang & Li, 2014).

AR uniquely would include a greater fidelity of real world environments. Team members could have the ability to interact face-to-face with bandwidth on multiple dimensions and the capacity to promote kinesthetic learning through physical movement in richly sensory spatial contexts (Dunleavy et al., 2009). This learning is an active and constructive process. For example, the CONNECT project employed AR and other technologies to develop a virtual science thematic park environment (Sotiriou & Bogner, 2008). The environment had two modes: the museum mode and school mode. Scenarios developed in the environment include both virtual and conventional field trips to science museums, pre- and post-visit curricular activities, and experiment and modeling activities. In this project, therefore, science learning at school was connected to learning experiences of the virtual and conventional museum visits, with the use of AR to augment students' visualization, experiments and models. An initial evaluation of the CONNECT project indicated that the environment positively influenced students' intrinsic motivation for learning science and conceptual understanding of the friction concept (Sotiriou & Bogner, 2008). A second aspect of this relates to the use of mobile devices in AR. With a wireless connection and location-registered technology, an AR system could enable collaborative and situational learning enhanced by virtual objects in real environments (Broll et al., 2008; Dunleavy et al., 2009). This would be useful for museums and other physical locations rich in content that is not easily displayed. Through mobile devices and geological positioning systems, learners have access to relevant information as they arrive at certain locations (Klopfer & Squire, 2008). The place or location for AR can be, for instance, a school campus where students actually study or the actual neighborhood where the school is located (Wu et al., 2013). Location-based AR has broad applicability in higher learning, allowing any number of applications where students study. Broad strolling campuses can transform their spaces with broad content while compact locations can open up areas with greater depth.

Universities can use actual or subjective representations in concert with the real world. They are often using subjective representations of reality. Their reality is augmented by linking information to prior knowledge, creating subjective mental representations. While constructivist theory provides the broader context of application of these technologies in education, social development theory provides a better correlation to these technologies.

Social Development Theory

The three major themes of social development theory describe how AR can support learning. Social interaction, the More Knowledgeable Other (MKO) and the Zone of Proximal Development (ZPD) are the foundations of social development theory that provide the framework for using AR in education. In addition, AR can support immersive learning, a major trend in education today. Immersive learning 'places individuals in an interactive learning environment, either physically or virtually, to replicate possible scenarios or to teach particular skills or techniques. Simulations, role play, and virtual learning environments can be considered immersive learning' (https://www.trainingindustry.com/glossary/immersive-learning/).

- Social Interaction plays a fundamental role in the process of cognitive development. In using partially immersive environments or socially augmented applications for real world interactions, learners are tapping into the affordance of social interaction to augment their understanding of the content. In formal and informal educational settings social interaction for adults also will assist in their cognitive development. Individual learners could use AR to interact virtually with each other during this stage. In group settings learners can have social interaction on a shared device, or have content displayed virtually, allowing group interaction across many barriers.
- The More Knowledgeable Other (MKO) refers to anyone who has a better understanding or a higher ability level than the learner with respect to a particular task, process or concept. The MKO is normally a teacher, coach or other adult, but the MKO could also be a peer, a younger person or even a computer (Social Development Theory, 2016). AR is used in the capacity to supply information or insight necessary for comprehension or knowledge to the learner. The ability to access relevant information side by side to real things or places is the MKO, with AR technology acting as the intermediary. AR can be used to provide information to the learner in real time, much like an educator would over the shoulder of a student in traditional classroom settings. The MKO is presented to the learner through their devices with content that supports the curriculum.
- The Zone of Proximal Development (ZPD) is the distance between a student's ability to perform a task under adult guidance and/or with peer collaboration and the student's ability to solve the problem independently. Considering that AR technology is acting as the MKO, or adult or peer guidance, ZPD is more efficient when the learner can solve problems independently with low risks. When the learner is utilizing AR to complete a task without guidance, technology can provide the role of the MKO in a classroom setting, potentially before the instructor engages, thereby decreasing the time needed to master a learning outcome.

Connectivism

AR is also aligned with Connectivism. Remaining under the broader Constructivist Theory, Connectivism explains how internet technologies have created new opportunities for people to learn and share information across the web and among the users. Connectivism enhances learning which can occur across peer networks online. Learning is a process that occurs within nebulous environments of shifting core elements—not entirely under the control of the individual. Most uses of AR technology addressed in this chapter reflect Connectivism, where learning is no longer an internal, individualistic activity.

In AR, learning occurs with new tools. Integrating augmented external sources can assist learners to master learning outcomes. Informal learning is a significant part of how we learn. Integrating technology into formal educational environments helps to integrate how learning occurs in a variety of ways in and outside of the classroom (Wang et al., 2011). AR can create collaboration opportunities, reducing the time needed to facilitate learning and creation in business and product development. AR can be used to emphasize human–product interaction and in product design and analysis. Educating new users in AR affords opportunities to collaborate with 3D renderings of future workspaces, as well as to educate the future workforce in the space.

Immersive Learning Theory and AR

Immersive learning has been used to digitally share content in virtual and modified worlds. This technology benefits from being based on code used for games like Second Life. Immersive 3D learning environments benefit from the detailed graphics capabilities and can provide the user some familiarity from online worlds (Wang et al., 2011). By using an interactive classroom capable of projecting virtual avatars within a physical classroom through augmented, virtual or immersive learning environments, educators can efficiently interact with virtual and physical students at the same time. While not every classroom can be modified to accommodate physical representations of virtual students located miles away, AR can provide realistic depictions of students when viewed through viewers. Social presence reflects the degree to which a person is perceived as 'real' in a mediated communication. Social presence theory is a seminal theory of the social effects of communication technology (Short et al., 1976). Utilizing AR with physical representations of people turning, looking and interacting with others in the classroom assists in creating a classroom environment that enhances interaction and the sharing of ideas.

These are among several factors that affect the level of presence, the extent to which a medium is perceived as sociable, warm, sensitive, personal or intimate when it is used to interact with other people. Virtual and Augmented Reality technology is about 'being there'. Presence is therefore partly to do with technology and partly to do with the user's state of mind.

AR and Mobile Learning

In recent years, mobile learning (mlearning), defined as 'anytime anywhere learning' and 'learning on the go' has transformed how people learn in many countries, especially in societies with a high degree of physical mobility, lack of access to high-speed internet and computers (Siemens, 2014). Mobile devices, especially smartphones, enable users to conveniently use their devices to interact with their environment. The intersection of virtual and physical realities spurred the birth of AR, which can elevate mobile learning experiences to newer levels (Wang & Shen, 2011). In addition, AR systems and apps also provide opportunities for higher-quality human interaction across the digital and real worlds. For learning, researchers perceived AR as a technology to connect theoretical knowledge to practical experience (Wang & Shen, 2011). Indeed, some studies show that AR can improve student performance in open-ended creative tasks such as urban and art design (Novak et al., 2012). In both studies, AR helped to simplify and facilitate the overall design process.

The fields of military, transportation and medical training are the early adopters of VR- and AR-supported training systems and they have discovered fascinating educational affordances provided by the technology, including improved user motivation and satisfaction (Wang & Shen, 2011). Higher-educational institutions around the world have been promoting the creation of virtual experiences and simulations, which can immerse participants in detailed digital environments enabling them to learn in novel and surprisingly ways (Wang & Shen, 2011; Wang et al., 2011). For instance, when Second Life first emerged, American universities were building their virtual campuses in there, with an aim to create virtual communities that can facilitate real community building and recruitment of prospective students. Some instructors were even teaching their classes in Second Life, by painstakingly creating avatars and instructional activities. The University of Essex, for instance, experimented with creating mixed realities by developing a learning environment that can blend virtual reality with the real world (Chen & Wang, 2008). Their goal is to better engage learners and also to promote interactions among students with diverse cultural backgrounds, so that they can collaborate and communicate more effectively. Years after, with the popularity of AR, instructional technology centers of some universities moved their focus to intelligent classrooms and Microsoft HoloLens. Nevertheless, all technology innovations need institutional and administrative support, as well as wider faculty participation. Without institutional support and user acceptance, most of these experiments become passing fads and non-sustainable.

It seems to us that the development and implementation of affordable AR-supported learning will be greatly affected by mobile learning and mobile technologies. Until now, mobile learning has stirred strong hopes in educators and scholars around the world but it has not become the mainstream modality of teaching and learning. Instead, it has been used successfully and widely to supplement teaching and learning activities. We believe that AR's future in education will very much depend on the further development of mobile learning.

Utilizing AR in Education

AR enables the 'ubiquitous computing' model and engaging anytime anywhere mobile learning. Students carrying mobile devices through real world contexts engage with virtual information superimposed on physical landscapes (such as a tree describing its botanical characteristics or a historic photograph offering a contrast with the present scene). This curated immersion can infuse digital resources throughout the real world, augmenting students' experiences and interactions (Callaghan et al., 2009). AR allows educators a level of flexibility which is useful in providing content. It can be utilized through a variety of personal devices. AR is portable and adaptable to a variety of scenarios and can be used to enhance content within a formal setting and provide context to content within both formal and informal settings. Content can be extended into the world and combined with other technologies to enrich their individual applications (Antonioli et al., 2014).

AR in and out of the Classroom

In formal training environments, classrooms are constrained to the resources available in the room. The learner's reach can be expanded through individual devices and internet connectivity to include readily accessible information, but discovering new data can be problematic as learners may easily lose focus without curated content to view. In AR, in contrast, material can be attached directly to any object in the learning environment (simulated or real). If a lesson is configured to allow mobile device access, the learner can retrieve stored content to augment the lesson from a variety of triggers. Aurasma, for example, is a free mobile application that can scan images to trigger content (e.g. videos, animations or 3D data) (Atherton et al., 2013). This capability allows educators to embed content virtually anywhere. 3D simulations displayed virtually over real objects can be interacted with and changed based on the learner inputs. Content can further be shared among peers and instructors.

As AR technology is more widely employed in education, a growing body of work offers evidence of its utility. In the European Union, Webel and colleagues (2011) undertook a study of VR and AR instruction for assembly and maintenance tasks, including the use of expert tele-consultation, 3D model overlays, haptic feedback, virtual notation and AR-based instruction (Van Krevelen & Poelman, 2010). Results have been favorable when comparing VR- and AR-based methods with classic instruction methods, including roughly equivalent transfer of training as compared with and a reduction of error when delivering AR instruction (Abu-Taieh & Abutayeh, 2009). In the defense sector, the US Navy (NAWCTSD/SPAWAR) is currently exploring the use of VR and AR for training and manufacturing under a Collaborative Research and Development Agreement (CRADA) with the Newport News Shipbuilding (NNS) for the CVN 78 Program. Scale models of crew compartments aboard CVN 78 with embedded AR functionality and user-triggered content have been successfully developed and demonstrated. The automotive industry has recognized similar utility for VR and AR, with Volkswagen and BMW employing the technology in their assembly lines (Earnshaw, 2014).

The popularity of games like Pokémon Go shows how AR can support mobile learning and gaming. A simple use of GPS tracking, data sharing and basic animation can harness a once popular card game to capture a new generation of users through a rudimentary use of AR. The popularity of the game helped the application develop a loyal following based on the premise of capturing hidden GPS-designated creatures to gain levels. This model for an entertainment-based game could be applied to mobile learning easily. Students could be tasked with finding historical points at a site that showcase significant events. Adult learners could use the GPS-based tracking and AR-based content to find a predetermined path through content in various settings. Brick and mortar institutions could use the basis for the game to augment their building with similar game outcomes, but instead of the goal to battle each other in 'gyms', the goal could be to collaborate and help each other master content that is relevant to skills appropriate for their curriculum.

AR content in formal settings can be attached to objects, pictures and sounds. Depending on the medium and capabilities of the software, learners can select to view the content again, providing learning materials that can be referred to after the instruction time has ended. Content that augments training materials can also be shared among peers and instructors. Classrooms that are more lab and student oriented can provide a more collaborative environment when sharing, and participation is extended beyond the physical limitations of the classroom.

If learners are encouraged to use their mobile devices for learning in the classroom, the learning environment will by design be extended from the traditional classroom to the outside when the student accesses that content again. Students preparing for a field trip to a museum can have the content related to that space provided through formal instruction and augmented content. When they are at the museum, they now can refer back to the content relayed to them at specific locations or access overlaid media about objects in the museum relevant to the lesson plan. Similar applications could be developed for conducting site visits in a myriad of settings from liberal arts to STEM training. AR technology can enable one to complete a tour of a campus or design a game to complete a tour of a facility. Completed individually or in group settings, students could move freely about the space that can foster a social constructivist approach in which students have independent roles and rely upon each other to complete the task (Dunleavy et al., 2009).

Through integration of AR into the classroom, students are more actively engaged in the learning process. Outside of traditional learning environments AR has many applications. Car manufactures can visualize systems for repair and troubleshooting, the medical community can see information about the body systems, and businesses can integrate AR for industrial processes or to see how the design of different components fit together. Volkswagen and BMW have already started to use AR technologies in their assembly lines (Earnshaw, 2014).

Limitations of AR

Although AR offers numerous possibilities for education, it has its limitations. As a case in point, a recent meta-analysis of VR and AR training in the automotive service industry concluded that although VR and AR education reduced error rates and was generally preferred over traditional training methods, research has focused almost exclusively on simple, skill-based tasks (Abu-Taieh & Abutayeh, 2009). Additional research is needed to establish the effectiveness of AR, and, in particular, methodologies for determining the proper application of AR for training complex, knowledge-based tasks. Further, research indicates that VR and AR training methods must be tailored to match the learner's expertise for maximum effect. AR can be misapplied, applied in excess or under-utilized if adopted without careful consideration of its target audience and the task (Van Krevelen & Poelman, 2010).

Wider Institutional Acceptance: Challenges, Barriers, Organizational Culture and Regulation

For any innovation to take off, there must be dedicated organizational support both financially and administratively, and also buy-ins from the users. For centrally managed educational systems, new innovations also need government support (Häfner et al., 2013). The development of online learning in the past decade exemplifies these perspectives. In both the U.S. and China, online learning addressed the shortage of brick and mortar universities, classrooms and instructors. For many universities in the U.S., online learning helped to greatly reduce expenses and to increase the number of course offerings, including the addition of summer semesters (Yu et al., 2005). In the U.S., degrees from programs offered by the same university, either obtained face-to-face, online or through a hybrid of both, do not differ in their qualities. In China, however, online learning programs, offered either by the open university system or by the online college of a particular university, often recruit students whose qualifications are not good enough for a traditional university. Therefore, degrees issued by online programs are not regarded as highly as degrees from traditional programs. This perception is deeply rooted in China's educational system and in its culture. Until these barriers are removed, online learning will not be the mainstream modality of teaching and learning (Tan et al., 2010).

Even though mobile learning is not considered a mainstream modality in education, its distinction from online learning has been diminishing. For instance, taking an online course on a laptop from a remote location is considered mobile learning. In many cases, mobile learning has been intertwined with online learning and has paralleled its development. From a user perspective, robust virtual classroom systems can effectively support teaching and learning and also provide ample opportunities for interaction. In recent years, the convenience, flexibility and mobility associated with online programs have greatly appealed to the mobile generation. Students can use an app such as Zoom or Google Hangout on their mobile devices to attend online meetings. Users do not draw a clear line between how they learn, either online or on their mobile device. They tap into all devices available and networks available to learn at work, at home, on the go or in a face-to-face classroom.

Universities such as Shanghai Jiaotong University (SJTU) and Colorado Technical University (CTU) have successfully implemented mobile learning, in combination with an innovative learning platform (Wang et al., 2011). The mobile learning system developed by the Online College of SJTU was used by tens of thousands of student and their instructors, in a variety of disciplines. This initiative was made possible by the strong leadership of the college's dean, a forward-thinking research and development team and a group of enthusiastic faculty. The CTU virtual campus provides students with access to multiple classroom delivery options, plenty of social networking opportunities and anytime communication. Mobile learning is enabled through an app called CTU Mobile™, with which students can access their virtual campus on their portable devices (http://www.ctucomm.com/). Many other universities, although they have not promoted mobile learning through a special brand like those of SJTU and CTU, use one of the learning management systems (Blackboard, Moodle, Canvas) that support mobile access. In light of the above discussions, we argue that mobile learning has become a dominant way of teaching and learning in North American universities. Mobile learning has not been clearly recognized as a modality of teaching due to the lack of widely accepted pedagogies and organizational promotion. In collaboration with a research team from the Open University of Shanghai and graduate students from San Diego State University, Minjuan Wang has published two models—Mobilegogy and the LTCS (Location, Technology, Content and Satisfaction) framework for designing instructionally sound activities (Machun et al., 2012; Wang et al., 2009; Wang et al., 2014). However, both models need to be tested with more users, faculty, instructional designers and students. We believe that the same model (see Figure 4.2) can apply to the design of mobile AR for formal and informal learning and have planned research to test this model in the near future.

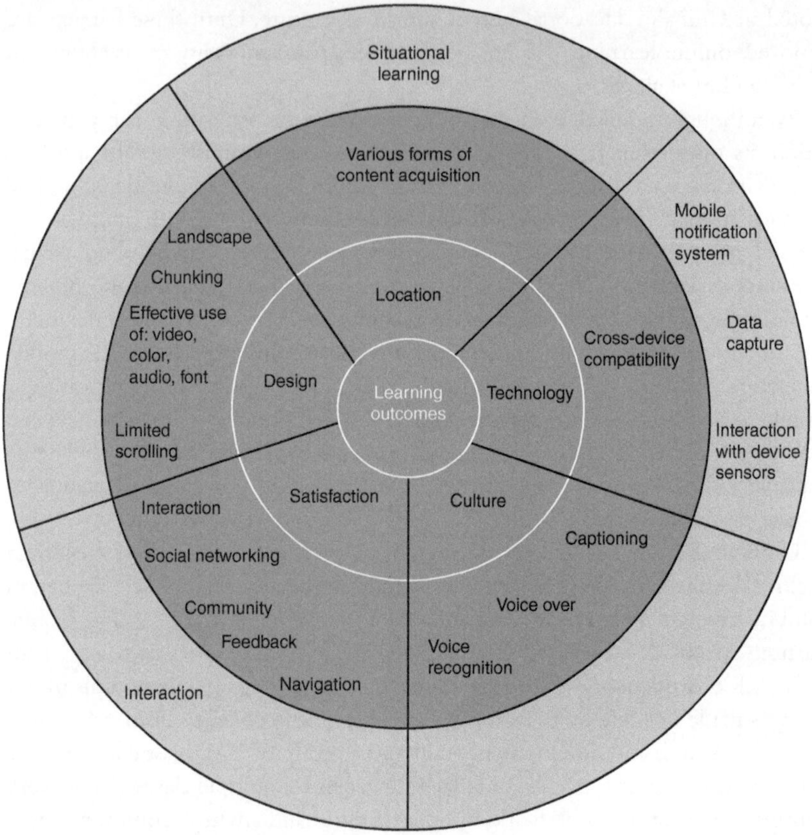

FIGURE 4.2 The Mobilegogy Model for Designing Effective Mobile Learning Activities (Machun et al., 2012).

In addition, the emergence and fast development of MOOCs in the last several years have taken away our attentions on mobile learning. Now that the hype around MOOCs has calmed down, it is time to revisit larger-scale implementation of mobile learning, especially when it can be coupled with AR technologies.

In countries like China, the development of mobile learning and VR has been market-driven. E-learning companies, supported by venture capital, are often at the forefront of new technology innovations. Some companies formed close relationships with higher-educational institutions and are contracted to develop technology-supported platforms and courses, such as learning management systems (LMS), VR applications and MOOCs. Even though mobile learning is still not recognized as an official modality for teaching or learning, China has a large number of mobile learners. In cities like Shanghai, subways are packed with mobile activities: reading, watching videos, playing games, online meeting or even doing homework. The popularity of WeChat, a social network app that blends Facebook, Twitter, WhatsApp, Instagram and more, gave mobile learning another strong boost. Learning platforms such as WeLearn have been developed inside WeChat and used for corporate training and for university teaching (Machun et al., 2012). Again, mobile learning has become a dominant way of informal learning, even though listening to a 'Sage on Stage' is the only accepted way to receive an acceptable degree.

For AR to further develop, it will have to follow the trajectory of mobile learning and to prove its value and necessity for learners. For mobile learning and mobile AR to break out from its relatively small niches into wider institutional acceptance, many efforts will need to work in concert, including: a) the establishment of a practical pedagogy, b) user appeal, c) organizational support to instructors and d) government support to universities in some countries. Nevertheless, the implementation of mobile learning has grown from small sparks to wild fires, and it indeed permeates the daily lives and workplaces of billions of people around the world.

Conclusion

AR is already in use by learners and is helping them to learn more efficiently as well as to increase their knowledge retention (Wang et al., 2015). New technology, as it is progressively introduced and used in more applications, with increased frequency, should be considered when integrating it into formal curriculum. AR activities should have design, hosting and platforms that ensure the content is accessible and relevant to the learner. It is an engaging way to interact in formal and informal learning environments and allows learners to collaborate. Through AR, the learner is provided with opportunities for deeper understanding of the subject matter. As results from the studies cited have shown, AR systems can provide motivating, entertaining and engaging environments conducive for learning. In addition, AR applications in educational settings are attractive, stimulating and exciting for students and provide cost-effective support for the users (Häfner et al., 2013). Technology has long been the barrier to implementation of AR and VR for educational outcomes. With recent developments and the use of mobile devices by virtually most of society, the bandwidth, technological capability and delivery method is in the hands of potential learners. When educators think of implementing this technology into their curriculum, the established theory exists to support its use, as well as to provide a framework for inclusion.

References

Abu-Taieh, E. M., & Abutayeh, J. M. (2009). Simulation environments as vocational and training tools. In Abu-Taieh, E. M., & El Sheikh, A. A. R (Eds.), *Handbook of research on discrete event simulation environments: Technologies and applications* (p. 15). Hershey, PA: IGI Global Publishing.

Antonioli, M., Blake, C., & Sparks, K. (2014). Augmented reality applications in education. *The Journal of Technology Studies*, 96–107.

Atherton, S., Javed, M., Webster, S. V., & Hemington-Gorse, S. (2013). Use of a mobile device app: a potential new tool for poster presentations and surgical education. *Journal of Visual Communication in Medicine, 36*(1–2), 6–10.

Broll, W., Lindt, I., Herbst, I., Ohlenburg, J., Braun, A. K., & Wetzel, R. (2008). Toward next-gen mobile AR games. Computer graphics and applications. *IEEE, 28*(4), 40–48. http://dx.doi.org/10.1109/MCG.2008.85.

Callaghan, V., Shen, R. M., & Wang, M. J. (2009). A mixed reality approach to hybrid learning in mixed culture environments. In J. Fong, R. Kwan, & F. L. Wang (Eds.), *Handbook of research on hybrid learning models: Advanced tools, technologies, and applications* (pp. 54–65). Hershey, PA: IGI Global Publishing.

Chang, M., & Li, Y. (2014). *Smart learning environments. Lecture notes in educational technology* (pp. 201–203). Heidelberg: Springer.

Chen, R., & Wang, X. (2008). An empirical study on tangible augmented reality learning space for design skill transfer. *Tsinghua Science & Technology, 13*(S1), 13–18. Tsinghua University Press.

Dunleavy, M., Dede, C., & Mitchell, R. (2009). Affordances and limitations of immersive participatory augmented reality simulations for teaching and learning. *Journal of Science Education and Technology, 18*(1), 7–22. http://dx.doi.org/10.1007/s10956-008-9119-1

Earnshaw, Rae A. (Ed.). (2014). *Virtual reality systems.* London: Academic Press.

Häfner, P., Häfner, V., & Ovtcharova, J. (2013). Teaching methodology for virtual reality practical course in engineering education. *Procedia Computer Science, 25*, 251–260.

Klopfer, E., & Squire, K. (2008). Environmental detectives: the development of an augmented reality platform for environmental simulations. *Educational Technology Research and Development, 56*(2), 203–228. http://dx.doi.org/10.1007/s11423-007-9037-6

Machun, P., Trau, C., Zaid, N., Wang, M. J., & Ng, J. (2012). MOOCs: Is there an app for that? *IEEE/WIC/ACM International Conferences on Web Intelligence and Intelligent Agent Technology, 3*, 321–325, Macau, China.

Milgram, P., & Kishino, F. (1994). A taxonomy of mixed reality visual displays. *IEICE Transactions on Information Systems, E77-D* (12), 1321–1329.

Novak, D., Wang, M. J., Callaghan, V., & Zhao, X. L. (2012). Looking in, looking out: A discussion of the educational affordances of current mobile augmented reality technologies. In J. Y. Jia (Ed.), *Educational stages and interactive learning: From kindergarten to workplace training* (pp. 92–106). Hershey, PA: IGI Publishing.

Short, J., Williams, E., & Christie, B. (1976). *The social psychology of telecommunications.* London: John Wiley & Sons.

Siemens, G. (2005). Connectivism: A learning theory for the digital age. *International Journal of Instructional Technology and Distance Learning, 2*(1), 3–10

Social Development Theory (Vygotsky). (2016, May 01). Retrieved from: http://www.learning-theories.com/vygotskys-social-learning-theory.html

Sotiriou, S., & Bogner, F. X. (2008). Visualizing the invisible: augmented reality as an innovative science education scheme. *Advanced Science Letters, 1*, 114–122. http://dx.doi.org/10.1166/asl.2008.012

Tan, L., Wang, M. J., & Xiao, J. (2010). Best practices in teaching online or hybrid courses: a synthesis of principles. In P. Tsang et al. (Eds.), *Lecture notes in computer science: Hybrid learning, 6248* (pp. 117–126). Berlin: Springer Publishing.

Training Industry Taxonomy. Immersive Learning. Retrieved September 18, 2017 from https://www.trainingindustry.com/taxonomy/i/immersive-learning.aspx

Van Krevelen, D. W. F., & Poelman, R. (2010). A survey of augmented reality technologies, applications and limitations. *International Journal of Virtual Reality, 9*(2), 1.

Wang, M. J., & Guo, M. (2015). A pilot study of using mobile platforms (WeChat and WeLearn) in college English curriculum. In J.C. Augusto & D. Preuveneers (Eds.), *Ambient intelligence and smart environments 19* (pp. 178–188). Prague: IOS Press.

Wang, M. J., Lawless-Reljic, S., Davies, M., & Callaghan, V. (2011). Social presence in immersive 3D virtual learning environments. *Advances in intelligent and soft computing, 92*, 59–67.

Wang, M. J., & Shen, R. M. (2011). Message design for mobile learning: Learning theories, human cognition, and design principles. *British Journal of Educational Technology, 43*(4), 561–575.

Wang, M. J., Shen, R. M., Novak, D., & Pan, X.Y. (2009). The impact of mobile learning on students' learning behaviours and performance: Report from a large blended classroom. *British Journal of Educational Technology, 40*(4), 673–695.

Wang, M. J., Xiao, J., & Chen, Y. (2014). Mobile learning design: The LTCS model. IE'14: *Proceedings of the 10th International Conference on Intelligent Environments*, 80–86, Shanghai, China: IEEE.

Wang, M. J., Zhang, H., & Saba, F. (2011). Instructional technology: Distance education and online learning. In X. X. Wang & R. H. Ouyang (Eds.), *Instructional technology: Critiques on the development of social sciences in the West* (pp. 166–186). Beijing: Renmin Publishing.

Webel, S., Bockholt, U., Engelke, T., Peveri, M., Olbrich, M., & Preusche, C. (2011). Augmented Reality training for assembly and maintenance skills. In *BIO Web of Conferences* (Vol. 1, p. 00097). EDP Sciences.

Wu, H. K., Lee, S. W.Y., Chang, H.Y., & Liang, J. C. (2013). Current status, opportunities and challenges of augmented reality in education. *Computers & Education, 62*, 41–49.

Yu, S. Q., Wang, M. J., & Che, H.Y. (2005). An exposition of the crucial issues in China's educational informatization. *Educational Technology Research and Development, 53*(4), 88–101.

Zap, N., & Code, J. (2016). Virtual and augmented reality as cognitive tools for learning. *EdMedia: World Conference on Educational Media and Technology, 2016*(1), 1340–1347.

5

IMPROVING ASSESSMENT AND FEEDBACK THROUGH VIRTUAL REALITY MOBILE LEARNING FOR HIGHER DEGREE APPRENTICES IN THE WORKPLACE

Katharine E. Jewitt

Introduction

The next few years are likely to see dramatic change across higher education (HE) given recent developments in the national context, such as the announcements about the Teaching Excellence Framework (TEF), including criteria relating to the quality of assessment and feedback; removing the caps on student numbers; increasing use and experimentation with new technologies; the potential introduction of the apprenticeship levy by the UK government (from April 2017, UK employers with a pay bill over £3 million each year will be required to invest in apprenticeships); the continued rollout of trailblazer apprenticeships (where employers design apprenticeship standards) and higher degree apprenticeships (where students combine work-based learning and training with a job, leading to a bachelor's and master's degree).

This chapter reports and reviews findings that emerged through a process of pause and reflection, asking whether assessment and feedback processes are sufficiently fit for purpose in this changing context, and what can we learn from recent innovation and initiatives in mobile learning?

The opportunity was taken to explore new processes and practices that underpin assessment and feedback practices in our institutions and across the wider HE sector for higher degree apprentices using virtual reality via mobile learning for informal learning, assessment and feedback practices in the workplace. It was identified that higher degree apprentices were suffering from a lack of confidence in relation to feedback and assessment and that students would be happier and more productive in the workplace if they were more confident about their job and subsequent assessment. We wanted a way to build confidence, improve skills and make learning, assessment and feedback in the workplace more enjoyable so apprentices would be encouraged to stay and complete their higher degree apprenticeship.

In a more competitive market, students will be seeking learning and assessment experiences that meet their specific and individual requirements. Learners' expectations of what constitutes a quality experience are also changing within the external consumer-oriented education landscape. Learner assessment and feedback becomes increasingly important within this landscape, and this presents a variety of new potential benefits for mobile learning, as well as bringing challenges for both students and institutions, together with employers and partners.

The study emerged through a process of wide consultation and included ideas generated from across UK educational institutions, including all categories of staff, students and a wide-ranging review of literature and liaison with a wide range of experts in industry and those employing apprentices for higher degrees.

There was an agreed focus for the study to:

a) Ensure that both informal and formal assessment and feedback is future facing, relevant and interdisciplinary;
b) Support work-based learning higher degree apprentices as active learners, with feedback and feed-forward, encouraging the building of productive, reflective learning and providing opportunities for dialogue; and
c) Ensure that assessment and feedback methods provide a rich, collaborative and engaged learning experience, which supports informal learning taking place in the workplace.

Role-play is a powerful tool for learning. However, in face-to-face classrooms or workplaces, introducing a role-play activity can make some students anxious. You may already have found that only some students are confident enough to engage in role-play. Virtual worlds provide collaborative spaces for learners to meet without these real-life limitations. They lend themselves well as a place to role-play, because learners can engage anonymously through the adoption of an avatar (their virtual presence) and learn from their peers through observation, discussion and peer assessment. Virtual reality accessed through mobile devices, provided a tool to offer engaging modes of interaction and help provide valuable, timely and just-in-time feedback on assessments. The main driver for using virtual reality via mobile devices was pedagogic. It was recognized that introducing mobile learning would provide flexibility, which would suit apprentices in the workplace and provide a more personalized and situated learning and assessment experience when not studying on campus.

There are many virtual worlds in existence for all ages, including Walt Disney's Club Penguin, Gaia Online for teens, Second Life® and popular virtual world video games like The Sims. This study used Second Life®, a free multi-user, three-dimensional, graphical virtual world, but the findings could be adapted to any other virtual world.

Theoretical Background

From the outset, the task was framed in terms of research from the FELTAG inquiry (Feltag, 2014), McLoughlin Report (McLoughlin, 2014) and '157 Group report: Further Education 2020' (Sherlock & Perry, 2013), which have highlighted key issues for institutions delivering apprenticeships. The research reports digital technology is not being widely used to change the nature of assessment, feedback and learning or radically enhance its accessibility. This is particularly the case for work-based learning and higher education apprentices where learning takes place informally, in the workplace, and would benefit greatly from mlearning.

Research has shown that learning is best supported 'when feedback is viewed as a relational process that takes place over time, is dialogic and is integral to the whole process of learning and teaching itself' (Sambell, 2011, p. 3). Feedback as dialogue means that the student not only receives initial feedback information, but also has the opportunity to engage the teacher in discussion about that feedback (Nicol & Macfarlane-Dick, 2006) and in context, defined by Brown (2010) as 'the formal or informal setting in which a situation occurs; it can include many aspects or dimensions, such as environment, social activity, goals or tasks of groups and individuals; time'. Good feedback is information that helps students to reduce the discrepancy between their intentions and the resulting effects (Nicol & Macfarlane-Dick, 2006). The use of mobile devices enables learners to be 'embedded in a realistic context at the same time as having access to supporting tools' (Futurelab, 2004) and can be used to 'facilitate, support, enhance and extend the reach of teaching and learning' (MoLeNET, 2010). Mobile learning provides a mechanism for contextualization of learning when it happens in the workplace. To make a difference, computers must be mobile and within 'arm's reach' (Soloway et al., 2001).

The New Media Consortium Horizon report for HE 2015 (The New Media Consortium, 2015) discusses the key issues likely to impact education the most over the next five years and contends that important developments in the next year will be with bring your own devices and flipped classrooms, and further ahead, with adaptive learning technologies, makerspaces, wearable technologies and the internet of things. There are a number of challenges associated with these technological developments in mobile technology: for example, redesigning learning spaces, personalized learning, improving digital literacy skills, the role of learning analytics in measuring learning, and the use of blended formal and informal learning and assessment. A key challenge in the use of mobile devices is to transform learning so it is a seamless part of the working day for the apprentice and 'not recognised as learning at all' (Futurelab, 2004).

O'Donovan, Rust and Price (2015) critically reviewed achievements in an organizational and institutional context in order to make recommendations guided through theory and practice. They suggest that our ongoing challenges and failings in the assessment and feedback space reflect the limited and fragmented impact of pedagogic research on practice. So it is both timely and appropriate to draw upon recent research and examine what we can do both individually and collectively to effectively share and embed the evidence from this research and the lessons learned from the many valuable projects (past and present) undertaken to improve assessment practice.

Despite the considerable research seeking improvements in assessment and feedback, evidence continues to demonstrate that the sector as a whole is failing in its assessment effectiveness, including attempts to address support, marking and feedback (O'Donovan, Rust & Price, 2015). The institutional appetite to support educational change in this area is driven largely by the National Student Survey data (HEFC, 2016) which has, since its inception, consistently indicated that students are notably less positive about assessment and feedback on their assignments than about other aspects of their learning experience (Williams & Kane, 2008), an assertion which still stands. JISC (a UK not-for-profit Joint Information Systems Committee supporting education and research) funded research in technology-focused assessment and feedback approaches with the end aim of a transformation in student experience (JISC, 2014). Mobile learning can be provided across and within contexts, providing a high level of personalization for the learners (Glahn et al., 2010).

There have been a number of significant research projects and partnering institutions attempting to confront the fundamental issues, for example, the Re-engineering Assessment Practices (REAP) project at the University of Strathclyde (2007) emphasized the importance of formative assessment and developing skills of self-regulation, while the Transforming the Experience of Students through Assessment (TESTA, 2016) and Programme-Assessment Strategies (University of Bradford, 2015) projects provided evidence of the importance of a holistic approach to assessment and feedback design and shared models to support the application of this approach across institutions. The Higher Education Academy paper 'A Marked Improvement' (The Higher Education Academy, 2012) provided institutions with a ready-made assessment review tool—which explicitly promoted the principle of assessment for learning, as well as the need for increased dialogue between staff and students around assessment—and even the QAA Quality code chapter B6 (QAA, 2013), talked of students' assessment literacy and the importance of dialogue with them about assessment.

There is a well-documented student demand for improved tutor–student–employer communication around assessment feedback in the workplace, as well as an understanding that student experiences of assessment feedback remain inadequately monitored.

Method

Case study methodology was adopted, described as 'a specific instance that is frequently designed to illustrate a more general principle' (Nisbet & Watt, 1984, p. 72) because as Robson (2002) describes, case studies examine real-life in order to provide a rich picture of the situation through the use of various data types. The case study is descriptive, explanatory and evaluative. Within this study, the case was 72 higher degree apprentices in STEM subjects working across four organizations in the UK using virtual reality via mobile laptop and tablet devices for a university term. A case study methodology was deemed advantageous because of taking place in a naturalistic environment where the researcher does not have control (Hitchcock & Hughes, 1995) and it reflecting the actual research situation, as much as is possible (Cohen et al., 2007). Triangulated evidence was utilized to address the potential weakness of observer bias in interpretation. The study integrated both qualitative and quantitative data to achieve depth and breadth in findings. Despite potential weaknesses, a case study provides a rich picture of the data required to examine assessment and feedback through virtual reality mobile learning. Data was collated through focus groups and interviews after the first week, in the middle of the term and at the end of the term.

There are many virtual learning spaces available in Second Life® which were utilized for the study, typically including a large lecture hall or auditorium for presentations and smaller classrooms for discussion. Managers and teaching staff held orientation sessions with their students, using Second Life® to learn how to control their avatar. Following the initial orientation sessions, exploration of the islands and scavenger hunts took place. Students then worked alone to explore locations, using the maps. Students were asked to visit points of interest, such as the Space Flight Museum and the Sistine Chapel.

A set of structured, collaborative role-plays were devised to explore challenging and common situations learners may face in their work, linked to their assessment. In these scenarios, learners took on a role and contributed and responded in character while in the workplace. After taking part in a few scenarios that were co-created between instructors and workplace staff, students then began to contribute to the assessment design and discuss what they wanted to spend more time on, in developing their skills and assessment practice. Apprentices worked in groups on assignments to assess and grade their own work, following set marking criteria. Apprentices had as much or as little support as they wanted while working in the virtual world, with more experienced staff on hand to provide advice and support.

Before starting, it was important to clarify the purposes of the role play. What learning was expected from the activity? Is this an opportunity for learners to develop skills or to develop empathy with another perspective? The purposes guided the design of each activity. For example:

- Student nurses practice communicating with different health professionals in a case conference.
- History students create alternative outcomes of a historical event by taking on roles of real-life figures to discuss, for example, the Cold War, the Cuban Missile Crisis.

- Contract law students explore alternative perspectives in a dispute by taking on the roles of different parties, e.g. architect, contractor, sub-contractor.
- Medical students take turns to take on the role of patient and doctor to build their confidence in having difficult conversations with patients.
- Language students meet in a museum and practice communication skills by taking on the role of museum guide and tourist.
- Trainee teachers rehearse behavior management strategies by facing common classroom scenarios.
- Law students perform a mock trial to develop mooting skills by presenting well-researched arguments to the 'judge'.

As with any role-playing, the debrief is important. Remind students about the purpose of the activity and give them time to reflect on what they have learnt. Encourage learners to reflect on how they can transfer skills learned to the real world. Peer assessment and self-assessment played a big part of the activity through the use of virtual reality via mobile devices.

The use of mobile learning enabled students to work with their employers on-site and with staff from the institutions working remotely to the workplace to de-construct assessments and deepen understanding. Taking on a role in a fictional situation is a useful way to provide an experience to learners and develop a more tacit understanding of course content and work-based processes. Staff and apprentices were set challenges and problems to solve which further bolstered their skills set and built confidence for assessments. Time was also scheduled for apprentices to come together at the end to socialize and chat about challenges and new discoveries. Social events were held for team building and tasks assigned to apprentices to discuss their course work and share problems. Designated times were set for virtual discussions, which provided apprentice/tutor/manager contact around specific topics, allowing apprentices to request and receive support from the right people around their employability skills development and study.

Findings

At first, we thought apprentices would not like the idea of being visible in the virtual world when learning, but we found that wasn't important and being able to access the virtual world at work and collaborate with more experienced staff members was.

Apprentices reported that they did not feel afraid to fail assessment tasks in the virtual world, unlike in the real world. It helped that they were able to repeatedly practice processes under the guidance of an expert to lead them to success, prior to assessments in the live work environment. Ninety-two per cent of the apprentices reported that in the real world they are often afraid to fail, but in the virtual world when they failed, it led to real satisfaction when they overcame the problem they'd first encountered.

> It's much easier to learn in Second Life® because you're disguised as an avatar, so the role-play is not embarrassing. It's a lot more fun than classroom training. It's safe to make mistakes and you can keep practicing as much as you need to. It's a lot more fun and interactive than real life.
>
> *Apprentice learner*

When learners are engaged in shaping and leading their own assessment and feedback this leads to benefits for all—learners, staff, the institution and the education system as a whole. Learners develop a greater sense of ownership over their learning and assessment, which leads to increased motivation, improved self-esteem and greater achievement.

Eighty-seven percent of apprentices reported that mobile learning allowed them more access to assessment opportunities through both their host institution and within the workplace, through easier access to communicate with peers, instructors and more experienced staff. It reduced the barriers previously encountered from working off campus. 78% reported they felt less isolated as a result of using mobile learning and 84% of the apprentices reported that their communication and feedback was a lot more personalized and immediate, due to the sharing of support across the organization together with support from the tutor.

> I found being able to enter the virtual world each morning and try out scenarios and receive immediate feedback from both my peers and from staff in the organization, as well as my tutor, greatly benefitted both my confidence and knowledge. If I got things wrong, I could try as many times as I needed to get it right.
>
> *Apprentice Learner*

The use of virtual reality has provided an unremitting focus on the use of feedback. Utilizing the software to host focus groups, we have used this as a mechanism for feedback to improve the assessment and feedback experience in the workplace. It has enabled effective strategies to keep all staff informed of the impact of feedback on management decision-making. Apprentices used activity within the virtual world to build a log of employability skills they acquired while studying, and managers could compare logs with their expectations and work requirements in reviews. The use of mobile learning provided a richer connection between staff and management, apprentices and the learning provider. The series of work-based learning scenarios built a portfolio of evidence for assessment and discussion.

> What I've liked best about mobile learning in the workplace is the increased opportunities for feedback and the endless chances to practice what we are learning. We can go into the virtual world whenever we want and there's always someone else working in there too, whether that is a peer or a

> member of staff. It really is anytime, anywhere learning. I really like the instant feedback mechanism both from individuals and in group discussion.
>
> *Apprentice Learner*

Skills acquired through the virtual activities transferred into the workplace. Staff and apprentices learned from each other through observing, mirroring, repeating and building confidence through being able to fail and try as many times as they needed before carrying out the work in the real-life workplace environment. The use of mobile learning provided a one-stop-shop for apprentices because everything was contained within the virtual world for apprentices to access—from assessment practice tasks to access to tutor feedback, a safe environment to make mistakes, work experience from staff and peer support. A key element which made the use of mobile learning so successful was its accessibility to everybody, whatever their ability and the mix of more experienced staff members supporting apprentice learners. The same experience could also take place on personal computers, which were on trolleys in the workplace, to support accessibility needs. Apprentices could enter the virtual world using their mobile device in the workplace to practice processes, ask questions and take part in discussion. Much as Amazon is the default platform for purchasing online, the virtual world became the 'go-to place' for students looking for a place to learn in a safe environment, where they could seek advice from more experienced staff. A spin-off of connected services began to emerge as the virtual world was used more, such as assessment opportunities, tuition and peer-to-peer support, evidence for e-portfolios and social, informal learning through chat.

A big issue standing in the way of mobile learning is organizational culture, and this issue is not to be underestimated. In this study, the continuous use of focus groups helped not only embed change but also continuously enhance the way mlearning was being used. Business fit is key, as well as ensuring buy-in from the senior management team. What helped mitigate issues and avoid the study failing was the focus on context. Time was spent upfront, understanding the issues needed to be addressed and working together in a three-way partnership between employer, institution and learners to design work-based learning activities within the virtual world to achieve objectives. It adopted a similar approach to trailblazer apprenticeships, where the employers worked as a group to ensure assessment activities were delivering the employability skills they were looking for. Apprentices regularly provided feedback on activities, and tweaks were made to tailor it better for learning and assessment.

Using mobile devices in the workplace enabled a lot more work activity to be incorporated into assessment tasks and closely related assessment criteria to what was taking place in the workplace. Marketing is also critical to make learners and staff aware of the opportunities available to them in ways which will attract them to try out new ways of learning and working.

Innovative approaches to communication through Second Life® via mobile devices have helped apprentices receive feedback on the quality of their work and make more informed choices about what to focus on next for their learning and assessment. It has provided a way for apprentices to talk directly with staff who they wouldn't normally come into contact with and to pose questions and make suggestions to which they receive timely responses. Senior managers also take part and are able to respond quickly to areas of concern. Learning providers and employers could informally observe, assess and provide feedback to apprentices and staff on their work and performance. Staff participants helped apprentices learn, demonstrated work to others, suggested responses to problems, explained how to do something, provided information, gave and received feedback, and worked with others collaboratively to aid assessment, feedback and feed-forward activities. These findings are also evident in the research by Cochrane (2010) which showed that when lecturers and students are interacting together and with others and integrating mobile learning into their many environments, students gain a strong sense of a learning community. Furthermore, he showed how learners were also critiquing and collaborating in peer assessment. Similar evidence is also seen from Kenny et al. (2009), who talk of a deeper, contextualized learning through mobile learning, which results in enhanced collaboration.

Conclusion

Most learning is informal and takes place at work under the guidance of non-educationalists. The use of mobile learning has allowed apprentices to work in ways they cannot in real life. There are so many opportunities in the use of virtual worlds via mobile devices, related particularly to a three-way partnership in learning and assessment between apprentices and employees, employers and learning providers: co-ordination of off- and on-the-job learning; real-time oversight for employers of their employees' progress; and use of virtual events at work to enrich learning, demonstrations of processes and development of learning communities. Its use is endless and hugely enriching by allowing learning to be much more opportunistic.

The use of virtual reality enabled apprentices and staff to learn any place and at any time. The training activities were developed to include examples drawn from everyday work activities so that they reflected real life. We used Second Life® as a way for apprentices to develop and practice skills. By taking on a role in a fictional situation, learners gain experience and develop a more tacit understanding of course content and work-based processes to improve assessment, feedback and feed-forward.

Entering the virtual world, learners can adopt a role to practice their skills via role-play activities, before transferring their skills to real life, prior to being assessed. Anonymity enables learners to build their confidence, make numerous attempts, develop and practice skills, take risks, fail and communicate more than they may in a real-life situation. There's no fear of asking a stupid question or saying the wrong thing, instead students can grow and develop in a safe environment. Virtual worlds have limitless possibilities in education.

This change is mirrored by a change in the role and significance of knowledge, learning, assessment and feedback. Connected universal mobile devices, the portal onto Web 2.0 services, change the nature of learning and knowing. Everyone with a mobile device can generate, store, share, discuss and consume images, ideas, information and opinions; can access the cloud and the services it provides; and can access each other. They can pursue, sustain or invent activities, assessment, feedback discussion specific to them, their location, their community and their history.

Institutions must respond to the unmanaged pressure from outside that comes from increased ownership, familiarity and expectations around universal personal technologies. Wider and more wholescale use of digitally enabled mobile learning is needed; one that balances ambitious top-down vision with a collaborative bottom-up responsibility for innovation. Increasingly, mobile learning happens outside institutions, while inside institutions students are forced to 'power down'. Now technology can be portrayed as socially transformative; technology 'is us' (Rettie, 2005). The first generation of technology enhanced learning was Web 1.0, with the web (and the educator) as broadcaster and students as readers and recipients; the second generation must be Web 2.0, where everyone is a writer and reader. Knowledge is now created locally, partially, contingently, for-me, for-now.

References

Brown, T. (2010). IDEO CEO Tim Brown: T-shaped stars: The backbone of IDEO's collaborative culture. Available from: http://chiefexecutive.net/ideo-ceo-tim-brown-t-shaped-stars-the-backbone-of-ideoae%E2%84%A2s-collaborative-culture

Cochrane, T. (2010). An mlearning journey: Mobile web 2.0 critical success factors in Montebello. In M.V., Camilleri, & A. Dingli (Eds.), *mLearn 2010: 10th world conference on mobile and contextual learning* (pp. 108–115). Valetta: University of Malta.

Cohen, L., Manion, L. & Morrison, K. (2007). *Research methods in education*. London: Routledge.

Feltag. (2014). *Further education learning technology action group recommendations*. Available from http://feltag.org.uk/wp-content/uploads/2012/01/FELTAG-REPORT-FINAL.pdf

Futurelab. (2004). *Mobile technologies and learning*. Available from http://admin.futurelab.org.uk/resources/publications-reports-articles/literature-reviews/Literature-Review203

Glahn, C., Borner, D. & Specht, M. (2010). Mobile informal learning. In Brown, E. *Education in the wild: contextual and location-based mobile learning in action*. Nottingham: University of Nottingham, Learning Sciences Research Institute.

HEFC. (2016). 2015 *National student survey summary data*. Available from http://www.hefce.ac.uk/lt/nss/results/2015

The Higher Education Academy. (2012). *A marked improvement*. Available from https://www.heacademy.ac.uk/system/files/a_marked_improvement.pdf

Hitchcock, G. & Hughes, D. (1995). *Research and the teacher*. (2nd ed.). London: Routledge.

JISC. (2014). Assessment and feedback. Available from https://www.jisc.ac.uk/rd/projects/assessment-and-feedback

Kenny, R.F., Park, C., Van Neste-Kenny, J.M.C., Burton, P.A. & Meiers, J. (2009). Using mobile learning to enhance the quality of nursing practice education. In M. Ally (Ed.), *Mobile learning: Transforming the delivery of education and training*. Edmonton: AU Press.

McLoughlin, F. (2014). *CAVTL Commission on adult vocational teaching and learning*. Available from http://repository.excellencegateway.org.uk/fedora/objects/eg:6878/datastreams/DOC/content

MoLeNET. (2010). *Mobile learning myths*. Available from http://www.molenet.org.uk/myths.html

The New Media Consortium. (2015). *The NMC horizon report: 2015 Higher education edition*. Available from http://cdn.nmc.org/media/2015-nmc-horizon-report-HE-EN.pdf

Nicol, D.J. & Macfarlane-Dick, D. (2006). Formative assessment and self-regulated learning: a model and seven principles of good feedback practice, *Studies in Higher Education*, *31*(2), 199–218.

Nisbet, J. & Watt, J. (1984). Case study. In J. Bell, T. Bush, A. Fox, J. Goodey, & S. Goulding (Eds.), *Conducting small-scale investigations in educational management*. London: Harper and Row.

O'Donovan, B., Rust, C. & Price, M. (2015). A scholarly approach to solving the feedback dilemma in practice. *Assessment & Evaluation in Higher Education*, *41*(6), 938–949. doi:10.1080/02602938.2015.1052774

QAA. (2013). *UK quality code for higher education – Chapter B6: Assessment of students and the recognition of prior learning*. Available from http://www.qaa.ac.uk/publications/information-and-guidance/uk-quality-code-for-higher-education-chapter-b6-assessment-of-students-and-the-recognition-of-prior-learning1#.V_DzQ_ArJjV

Rettie, R. M. (2005). Presence and embodiment in mobile phone communication. *PsychNology Journal*, *3*(1), 16–34.

Robson, C. (2002). *Real world research* (2nd ed.). Oxford: Blackwell.

Sambell, K. (2011). *Rethinking feedback in higher education: an assessment for learning perspective*. Bristol: The Higher Education Academy.

Sherlock, D. & Perry, N. (2013). *Further education in 2020: making the system work*. [Online]. Available from http://www.157group.co.uk/sites/default/files/documents/further_education_in_2020.pdf

Soloway, E., Norris, C., Blumenfeld, P., Fishman, B., Krajcik, J., & Marx, R. (2001). Log on education: Handheld devices are ready-at-hand. *Communications of the ACM*, *44*(6), 15–20.

TESTA. (2016). *Assessment Patterns that work*. Available from http://testa.ac.uk/index.php/resources/best-practice-guides/category/7-best-practice-guides

University of Bradford. (2015). *Programme assessment strategies*. Available from http://www.pass.brad.ac.uk

University of Strathclyde. (2007). *Re-engineering assessment practices in higher education*. Available from http://www.reap.ac.uk

Williams, J. & Kane, D. (2008). *Exploring the NSS: assessment and feedback issues*. York: Higher Education Academy.

6

A GLIMPSE INTO MOBILE LEARNING IN OPEN DISTANCE LEARNING (SOUTH AFRICA)

Thulile P. Shandu

Introduction

Mobile learning (mlearning) is associated with anytime, anywhere learning. In this chapter, learning without limits is viewed in juxtaposition with the flexibility afforded by Open Distance Learning (ODL). In ODL institutions such as the University of South Africa (UNISA), students continue with their daily routines while they pursue their studies. There is another distinguishing aspect of distance learning, however, which impedes, if not completely quells, chances of success: isolation. There is plenteous research on how ODL students can be supported cognitively and socially, as well as in administrative matters so as to minimize isolation, including through the use of mobile phones. A mobile phone is an integral part of students' daily lives as exemplified in the United States of America (Lepp, Barkley, & Karpinski, 2015), South Africa (North, Johnston, & Ophoff, 2014) and the Philippines (dela Pena-Bandalaria, 2007). This chapter, thus, explores how UNISA harnesses the ubiquity, versatility and availability of mobile phones for student support. The thrust of this paper is on using the experiences of students to develop guidelines for mlearning in ODL.

Mlearning at UNISA

UNISA researchers such as Makoe, Brown, Mbati, Letseka and Van Biljon, to name but a few, have conducted research in mlearning, specifically in Southern Africa, and their work is duly acknowledged. However, there remains a dearth of research on how mlearning manifests itself in ODL teaching and learning, and

TABLE 6.1 Prevalence of Mlearning at UNISA

Mlearning at UNISA	*Examples*	*Mlearning in Practice*
Institutional Policies	The Curriculum Policy of 2012 under 5.7.2 – pedagogy and technology	Embedding e-learning and mlearning in teaching and learning
Library	Mobile app for the library catalogue named m-oasis	Access to library resources such as searching the library catalogue, requesting materials and contacting librarians
Provisions	Agreements with computer companies as well as data service providers for discounted rates	Students can purchase bulk data at discounted prices Students can purchase tablets at discounted prices
Teaching and Learning	Some lecturers report using mlearning in their modules	SMSs containing definitions of key concepts sent to students
Administration	Registrations	Applications acknowledgement and updates through SMS
	Assessment administration	Updates on status of assignment (received, marked, returned) through SMS Examination and portfolio announcements through SMS Mobile app for submitting Multiple Choice Question (MCQ) assignments. Answers are entered and marked on the cellphone. The app provides a memo to each assignment
	Governance	National Student Representative Council (SRC) elections conducted through USSD codes on cellphones
	Announcements and information dissemination through SMS	Announcements of meetings and resources Management of ephemeral closures of university entrances

mlearning has not yet penetrated much in the pedagogic activities of institutions such as UNISA. Table 6.1 illustrates mlearning's prevalence at UNISA.

Table 6.1 portrays minimal learning associated with mobile phones, while Mobile Administration exists in different forms in what I term mAnnouncements, mVoting and mRegistration. There remains, therefore, a need for further investigation into the use of mobile phones for teaching and learning in ODL.

Background to the Study

The impetus for this study is two-pronged, based on Geduld's (2013) assertion that ODL students faced barriers including limited access to libraries and computers, high costs of computing, internet access, lack of English proficiency and poor writing skills. This study focuses on an intervention that could tackle the lack of English proficiency within a context marked by lack of access to computers. The study is influenced by the contention that without a developed vocabulary, students will struggle at the very least, and fail most likely (Wilkins, 1972). Research has shown positive results with the use of mlearning (Thornton & Houser, 2001; Hodgkinson-Williams & Ng'ambi, 2009); therefore, this study focuses on how ODL students learn with mobile devices.

Theoretical Framework

This research was framed within the Social, Cognitive and Teaching presences of the Community of Inquiry (CoI). The CoI framework affirms that it is through these presences that deep and meaningful learning is facilitated in online and, in this study, mobile learning contexts (Anderson, 2016). Social presence refers to interaction for teaching and learning at a distance (Anderson, 2016). Interaction pertains to student and student; student and teacher; student and content; and, for this particular study, student and device (Makoe, 2012).

Social presence is necessary for the development of cognitive presence (Garrison & Arbaugh, 2007). Cognitive presence is "the extent to which learners are able to construct and confirm meaning through sustained reflection and discourse" (Garrison & Arbaugh, 2007, p. 159). According to Garrison, Anderson and Archer (2001), cognitive presence is accomplished when a problem is identified for further inquiry (triggering event) and students explore the issue, both individually and corporately, through critical reflection and discourse (exploration). They then construct meaning from the ideas developed during exploration (integration), and finally learners reach a higher level thinking in developing their ideas (resolution). Garrison, Anderson and Archer (2001) contend that the integration phase typically requires enhanced teaching presence towards resolution. Teaching presence refers to "the design, facilitation, and direction of cognitive and social processes for the purpose of realizing personally meaningful and educationally worthwhile learning outcomes" (Garrison & Arbaugh, 2007, p. 163). The interrelatedness of the presences is evident in their definitions in that it is in the social presence that teaching presence facilitates cognitive presence.

The CoI was fit for this study because it informs both research and practice (Anderson, 2016); provides templates as it aids practitioners with instructional design models (Wicks & Sallee, 2016); and is a tested, validated and well-researched theoretical framework (Garrison, Anderson & Archer, 2010).

Research Aims

This study aimed to investigate ODL students' experiences of using mobile technologies to enhance their vocabulary.

Objectives:

1. Examine how ODL students use mobile technologies to enhance their vocabulary.
2. Explore ODL students' experiences of mobile-based vocabulary activities.
3. Suggest guidelines for the use of mobile technologies to enhance students' vocabulary in ODL.

Methodology

This study used Design-Based Research (DBR) to respond to the research aim. DBR was deemed appropriate because it allows for insight into how, when and why educational innovations work in practice (Baumgartner et al., 2003). DBR is often based on collaboration among researchers and practitioners in real-world settings (Wang & Hannafin, 2005) and in this research I took on the role of practitioner and researcher because the teacher can also be researcher (Opie, 2004). DBR "produces useful products (e.g., educational materials) and accompanying scientific insights into how these products can be used in education" (Bakker & van Eerde, 2015, p. 430) and this study produced 1) a vocabulary learning programme that includes the use of a vocabulary app together with WhatsApp and 2) principles to guide mlearning for vocabulary development in ODL.

In order to detail how DBR was used in this study, a generic model for design research (GMDR) by McKenney and Reeves (2014) is used as an outline: analysis and exploration; design and construction; and evaluation and reflection. According to the iterative nature of DBR (Baumgartner et al., 2003), this study is in its third cycle. Using GMDR as an outline, the focus of this chapter is situated in phase three of the cycle: evaluation and reflection. Table 6.2 is a summary of the phases of the study thus far.

Methods

This study collected data through observations, one-on-one text-based virtual interviews, as well as text-based virtual focus groups using online threaded discussions within WhatsApp groups to explore the research questions. The use of virtual focus groups is deemed theoretically sound by Turney and Pocknee (2005) who concluded that such data collection was related to Krueger's (1988) criteria for focus groups, which include involving people; being conducted in a series; having relatively homogenous participants; being a method of data collection; collecting qualitative data; and constituting a focused discussion. James

TABLE 6.2 Iterative Phases of the Study

Exploration	*Design and Implementation*	*Evaluation and Reflection*
Iteration 1: Teaching and Learning on VocUp		
Literature review on mlearning and vocabulary learning Literature on design of the vocabulary mobile app, structure of the vocabulary course, including catering for interaction, and instructional approaches for vocabulary teaching	Created Word Capsules Designed and developed VocUp Disseminated to colleagues Disseminated to participants	Comments on VocUp: Refining Word Capsules Feedback on pronunciation clips, example sentences and exercises VocUp crashes
ITERATION 2: Teaching and Learning on WhatsApp		
Literature review on mlearning and vocabulary learning Observations of group interaction on WhatsApp	Refinement of vocabulary programme Used feedback from first iteration to revise Word Capsules (included more examples, pronunciation sound clips and solicited examples from participants) Sent Word Capsules on WhatsApp	Focus group interviews Observations Found that some participants still preferred to use WhatsApp while others preferred VocUp
ITERATION 3: Teaching and Learning on WhatsApp and VocUp		
Literature review on mlearning and vocabulary development Interviews with participants	Further refinement Used feedback from second iteration to improve teaching on WhatsApp (eliminated teaching similar sounding words together), improved activities (added longer writing activities) and encouraged more interaction through increased prompting	Interviews on the use of WhatsApp and/or VocUp Observations of focus group

and Busher (2012) list a number of benefits associated with virtual focus groups, including time and cost effectiveness (no transcription), and accuracy of data for the researcher, as well as convenience for the respondents who are able to participate in conversations without travelling to interview venues. Shapka, Domene, Khan and Yang (2016) compared face-to-face focus groups to virtual focus groups and concluded that, even though virtual groups produced fewer words, data quality was not affected by using either of the two methods. In this study, the benefits of text-based virtual focus groups were amplified by mobility because the interviews were conducted on the mobile phone where participants could respond whenever and wherever they found an opportunity to do so.

Mobile App Development

In preparation for the study, I developed a vocabulary mobile app called VocUp. VocUp sends daily lessons in the form of what I termed "Word Capsules". Each Word Capsule contains the word of the day, part of speech, definition, three examples illustrating the appropriate use of the word and three exercises to assess the user's grasp of the word. The vocabulary items were selected from both versions of the Vocabulary Levels Test (Nation & Waring, 1997; Schmitt, Schmitt & Clapham, 2001). According to Nation (1990) students who operate at the 10,000-word level are able to read widely and cope with specialized subject area study. I used vocabulary from the key as well as distractors for the Word Capsules. VocUp updates at 5:00 every morning with a notification envelope so users can start learning new words as soon as they are ready. By sliding between screens, users can choose to focus on the word of the day, example sentences or practice exercises. Figure 6.1 presents screenshots of VocUp on a mobile phone.

A link was sent to colleagues, supervisors and relatives who were students at UNISA for testing purposes. Revisions were made four times as I received feedback on the app interface as well as the vocabulary content through informal chats over the phone and WhatsApp. Feedback saturation was reached when testers said there was nothing more they wanted to change.

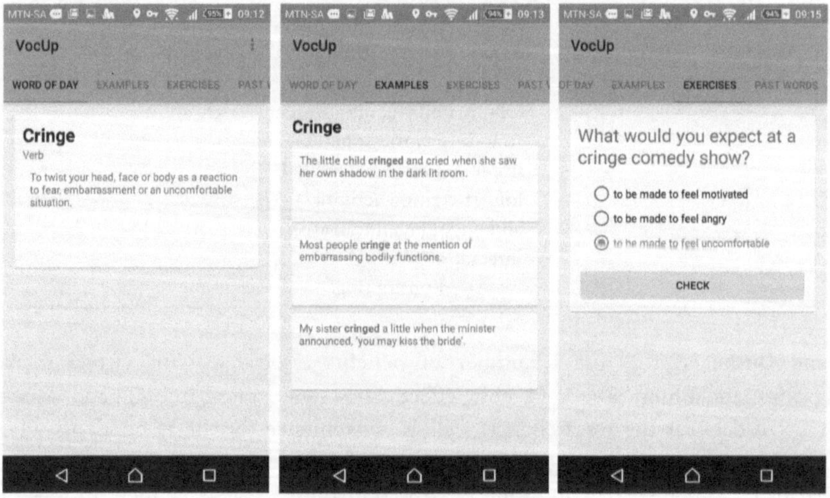

FIGURE 6.1 VocUp Screens on Mobile Phone.

Participants

The participants were selected from first-year students at UNISA through purposive sampling. The strength of purposive sampling lies in its "intentional bias" (Tongco, 2007, p. 154). I sent electronic invitations to students registered for a module I was teaching. The invitations were posted on the LMS's Announcements tool, which automatically sends emails to students alerting them of new announcements. After three reminders for the invitation yielded no response, I emailed the invitations to an academic literacy facilitator at one of UNISA's learning centres who distributed the invitations to participants during one of his workshops. The interested students signed consent forms and left them with the facilitator. When I collected the consent forms, I invited more students who were attending another workshop. I received a total of 29 signed consent forms from students who were all registered for first-year English for Academic Purposes who would provide insight into the experiences of first-year students. The sampling was still purposive, with a difference in the mode of invitation, from electronic to face-to-face invitation.

Method

The 29 participants were assigned to four WhatsApp groups of six members, with the fifth allocated five members. I numbered the participants to 29 and counted to six consecutively four times, and the fifth time counted to five. After an orientation on WhatsApp, which included agreements on mutual respect and encouragement for as much participation as possible, I sent the VocUp link to participants with instructions for downloading the app. I then sent the first interview questions to the groups after they had downloaded the app. The app crashed on the second day, owing to debugging errors. We agreed with the participants that the Word Capsule activities would be sent on WhatsApp while VocUp was being fixed. Four weeks into the study, I sent another set of interview questions to the

TABLE 6.3 Methodology at a Glance

Date (2016)	*Research Activity*	*Methodology Activity*
February	Workshop on app development	Literature review
February–May	Writing Word Capsules App development and pilot	Literature review
19 May	Orientation and VocUp download: first word sent Activities over app only	Observations
20 May	VocUp crash	Observations First interviews
21 May–20 June	Activities over WhatsApp	Observations Second interviews
21 June–18 August	App and WhatsApp	Observations
18 August	Study ends	Final interviews

groups as well as to individuals. The final set of questions was sent at the end of the study. Table 6.3 summarizes the research methods in this study.

Analysis and Discussion

Data analysis followed a hybrid approach that was both framework-driven deductive and data-driven inductive thematic analysis (TA) similar to that of Fereday and Muir-Cochrane (2006). For deductive coding, the CoI was used as both a framework for grounding and validating the themes presented and as an organizational apparatus towards a more lucid presentation of themes. The CoI was chosen as a theoretical template because it accounts for all the major themes of successful online courses (Anderson, 2016); it is widely used as a framework (Garrison, Anderson & Archer, 2010); and the presences have been confirmed through factor analyses (Garrison, Cleveland-Innes & Fung, 2004). For inductive coding, I followed Braun and Clarke's (2006) six-step thematic analysis (TA): become familiar with the data, generate initial code, search for themes, review themes, define and name themes and then, finally, produce the report.

Of note in Table 6.4 is the reflective nature of TA. The themes from the first group in WhatsApp were compared and refined with those in other WhatsApp groups and subsequent focus group interviews. These were then compared and validated with one-on-one interviews. The researcher's observation notes were also used to refine and validate themes. As illustrated in Table 6.4, the 25 initial codes were refined into seven themes, which were then merged into the existing CoI framework: *social presence* (community bonds made); *cognitive presence* (negotiated interaction, learning vocabulary was facilitated through mlearning); *teaching presence* (the facilitator is not replaceable). The three themes from inductive coding (*the mobile phone is a source of benefits and challenges*; *personal issues affect learning*; *things to change*) that did not seem to fit in with the framework were not discarded because each theme tells a story (Braun & Clarke, 2006, p. 92) and "it is important to consider how it fits into the broader overall 'story' that you are telling about your data, in relation to the research question".

Social Presence

> *7/19/16, 13:05 – P16: Good news I passed my English even though i ran out of time, I did not finish remember?*

Social presence, provides insight into how participants project themselves as real people (Garrison, Anderson & Archer, 2000). In this study, the participants were able to express themselves in an environment that was a socio-emotional outlet and a forum that fostered group cohesion for learning. These activities are matched with CoI's subcategories of open communication, affective expression and group cohesion. Open communication was evident in the participants' freedom to use

TABLE 6.4 Thematic Analysis Steps Based on Braun and Clarke (2006)

Thematic Analysis Step	*Activity*	*My Activity*	*Number of Themes*
1. Become familiar with the data	Transcription and finding meaning through repeated reading of the data	I read and repeatedly read through the hundreds of WhatsApp posts and interview responses as well as my notes	Over 1,000 posts on WhatsApp
2. Generate initial code	Finding initial coding of interesting features from raw data	I highlighted comments in different colours, with similar comments in similar colours	25 highlighted codes
3. Search for themes	Finding common themes from the initial codes	I grouped my themes in a table and looked for relationships and commonalities between the initial codes	17 themes
4. Review themes	Reduction of data and codes	I merged related themes	Seven reviewed themes
5. Define and name themes	Refining themes so they demonstrate the overall story the analysis tells	I refined the themes based on the study objectives. I then subdivided my themes into the CoI elements and those that did not fit into the CoI elements	Three themes related to CoI elements Two additional elements as pertaining to this study
6. Producing the report	Reporting is the presentation of the analysis, together with extracts from data. Reporting also involves consideration of the research objectives as well as relevant literature	I had to integrate findings and literature in relation to the research objectives. I used CoI to provide guidelines since it is the theoretical framework for this study	The analysis chapter

a language they felt comfortable with. In multilingual settings, such as South Africa with 11 official languages, it was important that language learners used a language they were familiar with. The researcher, being multilingual herself, did not struggle with engaging in discussions. The rare occasions where a word was not understood in a language other than English, other group members helped to define it in English. Other languages were also used to define English concepts.

Openness was noted in instances of affective expression where group members encouraged others when they wanted to give up; celebrated the achievements of others; and admonished each other when they were not contributing to discussions. Group members also used emoticons to express their emotions in the absence of non-verbal cues associated with face-to-face communication. The use of emoticons supplements text in online communication (Rezabek & Cochenour, 1998) for participants to express varied emotions.

The socio-emotional aspects of the social presence are important, yet "not sufficient for educational purposes" (Garrison & Arbaugh, 2007, p. 161). It is in group cohesion that participants build knowledge together.

Cognitive Presence

> *6/18/16, 10:05 – P24: Sometimes you don't understand particular word then someone come with differet ideas and evin what that word means in your mother tongue.*

Participants recognized "that they are not there for purely social reasons" (Garrison, 2007, p. 159), but through social interaction, participants gained cognitive benefits. The cognitive presence develops through phases of a triggering event where some issue or problem is identified for further inquiry; exploration, where the issue is explored; integration, where learners construct meaning from the ideas developed during exploration and then resolution (Garrison, Anderson, & Archer, 2001). The study experienced these phases, one of which was when participants were trying to make sense of "vie" as the word of the day. The triggering event was an incorrect explanation of vie in a language other than English. The other participants joined the conversation as they tried to clarify the meaning in various languages and examples. Exploration was facilitated by the researcher who asked questions and shared examples. Integration was reached when participants shared their own real-life examples. In the end, there was agreement on the meaning of "vie". The overarching observation is that cognitive benefits were harnessed through interaction.

Teaching Presence

> *8/20/16, 08:03 – P3: It was super convenient. No difficulty. It was very well presented, easy to learn*

Cognitive presence is facilitated through social presence, but there is a need for a facilitator to design, facilitate and provide direction (Garrison, Anderson, & Archer, 2001). It was noted in the study that the willingness and competence of the facilitator (also researcher) contributed to the effectiveness of the lessons. While the participants appreciated the attention they got from the facilitator, it was mostly the contribution to their learning that was lauded. When one of the participants called the facilitator a language connoisseur (a noun that had been taught prior), it was a sign that the new vocabulary was being learned and used, but it was also evidence that participants acknowledged the competence of the facilitator. Being a non-native speaker of English contributed to her teaching method; she was able to share her own language-learning strategies towards achieving proficiency in English. It was also observed that while the definitions and example sentences were deemed helpful by participants, it was the exercises that prompted much discussion and helped the participants to fully grasp vocabulary. The participants confirmed that they loved doing exercises as they tested their understanding. The two participants who preferred working only on VocUp also emphasized that the exercises helped them work on their own since feedback was immediate on incorrect options (with the option shaking and turning red) and correct options (turning green). It took the teaching expertise of the facilitator to design activities and exercises that would facilitate the cognitive presence. Mlearning, therefore, requires an array of competencies from and full participation of the teachers (Makoe, 2012).

Themes Outside of the CoI

There were three themes that did not fall into the three elements of CoI, but which are crucial as they also tell the story of the experiences of participants.

First, the technology issues immanent in mlearning in ODL were too prominent to be ignored. Data cost, connectivity and phone compatibility were cited as some of the reasons the participants could not fully participate in activities. Responses such as *5/19/16, 17:33 – P3 I can't download this app* prompted further inquiry, and reasons for struggling included the device itself, internet connectivity, financial issues and security issues—some participants were wary of downloading VocUp for fear of malware. The threat of malware is so real that Kambourakis posits that system and data security and privacy are "crucial for any educational realm" (2013, p. 69). These were also some of the reasons put forward for preferring WhatsApp, which was deemed safe, inexpensive and trustworthy, based on familiarity. The fact that the app crashed around the beginning of the study elucidated the lack of trust even after assurances that it was fixed and safe.

Second, there were personal issues that were barriers to full participation. Financial and time constraints were the predominant issues. While participants appreciated the vocabulary activities, they admitted to time constraints related to work commitments, studies, family responsibility, sickness and travel. These were explicitly mentioned and they were observed when some group members suddenly became less active in the group. Comments such as these were common: *5/28/16, 11:28 – P2: I will check but I'm still at work* and *5/31/16, 10:18 – P6: I will respond later am buizy for now, and I see this is interesting.* Even the most exciting discussions and informative lessons could not be attended to due to time constraints.

Finally, the participants were vocal about what changes they wanted implemented for vocabulary learning. These ranged from content (including more exercises) to facilitation (removing members who were not contributing to discussions), as well as institutional (ensuring that more modules incorporated mlearning in the curriculum). Some of these suggestions can be incorporated in the revised versions of VocUp while others would need institutional changes, but what is crucial is that students in ODL have a say in how they should be taught; blindly teaching without acknowledging the needs and insights of students is not sustainable and undermines their intelligence and the role students should play in their learning.

Implications

> *5/30/16, 07:38 – P5: I never bothered myself to take something to eat on lunch time. It was my lunch. My friends notice that am not interested to their conversation anymore. I learned new words. Thanks pearl*

CoI provides needed guidance and insight regarding online learning in ODL, but there is a need for its adaptation so it acknowledges the unique characteristics of the ODL context. While cognitive and teaching presences play a crucial role, the social presence takes centre stage since it is the space in and through which the presences operate. This study further concludes that device and learner presences should feature prominently in any consideration for effective teaching and learning in ODL. No matter how well-designed and well-facilitated lessons are, no matter how open the social presence, the device can be a barrier to those resources. Furthermore, if the learner is not prominently considered, then the best designed interventions can be deemed ineffective. Figure 6.2 is a suggested interpretation of the CoI for this context.

Because DBR prescribes the production of a product as well as principles guiding the use of that product (Bakker & van Eerde, 2015), this study produced a pool of vocabulary learning lessons which can be used as a mobile app for students who prefer to study on their own or with WhatsApp to facilitate interaction. The principles guiding the use of mlearning technologies in ODL relate to convenience, trust and the role of the students:

1. Students in ODL have many other responsibilities vying for their attention in the form of time and money. They steal pockets of time to focus on learning, and teachers should take care that they design convenient interventions so that those moments can be used effectively.
2. Trust is an important aspect in ODL where the geographical distance is a marked characteristic. If students cannot trust the competence of the teachers or the interventions they present, then learning is hindered. This calls for greater care in planning, designing and facilitating interventions.
3. Finally, students in ODL bring life experiences and expectations that should not be undermined or overlooked. There has to be a way of incorporating the students' preferences or, at the very least, acknowledging their contexts and prior knowledge.

Conclusion

Through a qualitative inquiry, the experiences of students with regard to learning vocabulary through mlearning technologies were investigated. The study

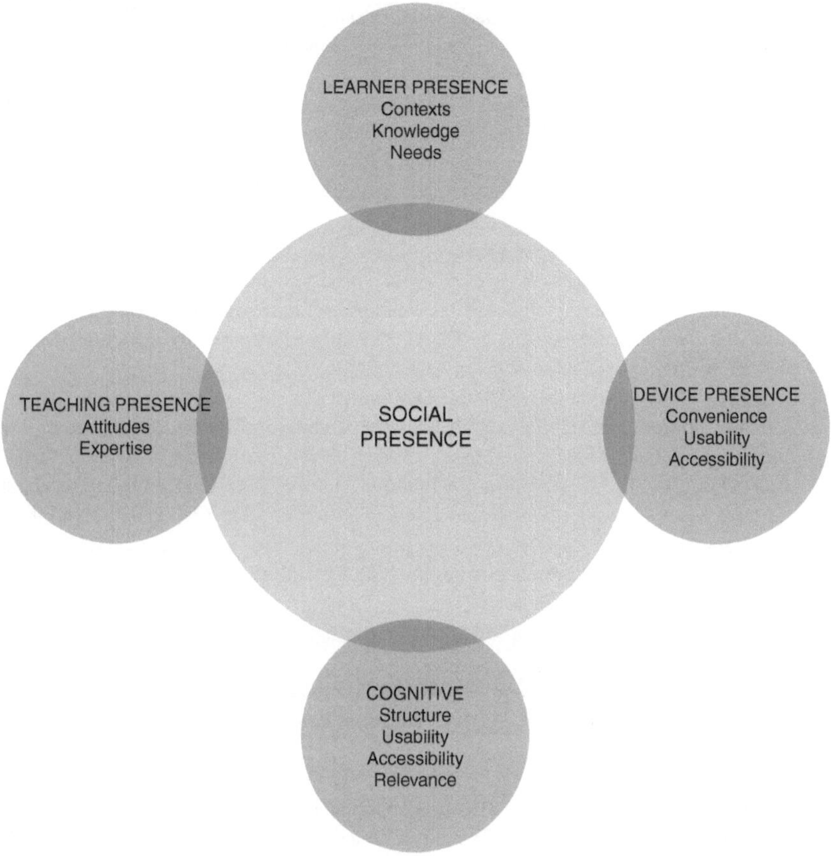

FIGURE 6.2 Expanded Community of Inquiry. Adapted from Garrison, Anderson and Archer (2001).

highlighted the importance of convenience with regard to technology since ODL students face time as well as financial constraints. For learning vocabulary, ODL students find engaging with others effective while vocabulary exercises were a useful tool for testing understanding and prompting interaction. The guiding principles to mlearning in ODL highlight convenience and trust, as well as acknowledgement of the students. All these facets work towards ensuring that the student, no matter how isolated, is able to fully interact with fellow students and teachers, as well as study content towards effective learning.

References

Anderson, T. (2016). A fourth presence for the community of inquiry model? Virtual canuck. Retrieved from http://virtualcanuck.ca/2016/01/04/a-fourth-presence-for-the-community-of-inquiry-model/

Bakker, A., & van Eerde, D. (2015). An introduction to design-based research with an example from statistics education. In Bikner-Ahsbahs, A., Knipping, C., & Presmeg, N.

(Eds.), *Approaches to qualitative research in mathematics education*, 429–466. Dordrecht, The Netherlands: Springer. Retrieved from http://link.springer.com/chapter/10.1007/978-94-017-9181-6_16#page-1

Baumgartner, E., Bell, P., Brophy, P., Hoadley, C., His, C., Joseph, D., Orrill, C.H., Puntambekar, S., Sandoval, W.A., & Tabak, I. (2003). Design-based research: An emerging paradigm for educational inquiry. *Educational Researcher, 32*(1), 5–8.

Braun, V., & Clarke, V. (2006). Using thematic analysis in psychology. *Qualitative Research in Psychology, 3*(2), 77–101. Retrieved from http://www.tandfonline.com/doi/abs/10.1191/1478088706qp063oa

dela Pena-Bandalaria, M. (2007). Impact of ICTs on open and distance learning in a developing country setting: the Philippine experience. *International Review of Research in Open and Distance Learning, 8*(1), 1–15. Retrieved from http://www.irrodl.org/index.php/irrodl/article/view/334/793

Fereday, J., & Muir-Cochrane, E. (2006). Demonstrating rigour using thematic analysis: A hybrid approach of inductive and deductive coding and theme development. *The International Journal of Qualitative Methods, 5*(1), 1–11.

Garrison, D.R. (2007). Online community of inquiry review: Social, cognitive, and teaching presence issues. *Journal of Asynchronous Learning Networks, 11*(1), 61–72. Retrieved from http://www.tonybates.ca/2008/08/07/online-community-of-inquiry-review-social-cognitive-and-teaching-presence-issues/#sthash.3ZuzqKzS.dpuf

Garrison, D. R., Anderson, T., & Archer, W. (2000). Critical inquiry in a text-based environment: Computer conferencing in higher education. *Internet and Higher Education, 2*(3), 87–105.

Garrison, D. R., Anderson, T., & Archer, W. (2001). Critical thinking, cognitive presence, and computer conferencing in distance education. *American Journal of Distance Education, 15*(1), 7–23.

Garrison, D. R., Anderson, T., & Archer, W. (2010). The first decade of the community of inquiry framework: A retrospective. *The internet and higher education, 13*(1), 5–9.

Garrison, D. R., & Arbaugh, J. B. (2007). Researching the community of inquiry framework: Review, issues, and future directions. *The Internet and Higher Education, 10*(3), 157–172.

Garrison, D. R., Cleveland-Innes, M., & Fung, T. (2004). Student role adjustment in online communities of inquiry: Model and instrument validation. *Journal of Asynchronous Learning Networks, 8*(2), 61–74.

Geduld, B. (2013). Students' experiences of demands and challenges in open distance education: A South African case. *Progressio, 35*(2), 102–125.

Hodgkinson-Williams, C., & Ng'ambi, D. (2009). Case study 5: Mobile learning. *Opening scholarship*. University of Cape Town: Centre of Educational Technology. Retrieved from http://www.researchgate.net/profile/Cheryl_Hodgkinson-Williams/publication/228788595_Case_study_5_Mobile_learning/links/00b495159da73c3476000000.pdf

James, N., & Busher, H. (2012). Internet interviewing. In J.F. Gubrium, J.A. Holstein, A.B. Marvasti, & K.D. McKinny (Eds.), *The SAGE handbook of interview research: The complexity of the craft* (2nd ed.) (pp. 177–188). Thousand Oaks, CA: Sage Publications Inc.

Kambourakis, G. (2013). Security and privacy in m-learning and beyond: Challenges and state-of-the-art. *International Journal of u-and e-Service, Science and Technology, 6*(3), 67–84.

Krueger, R. A. (1988). *Focus Groups: A practical guide for applied research*. UK: Sage.

Lepp, A., Barkley, J. E., & Karpinski, A. C. (2015). The relationship between cell phone use and academic performance in a sample of US college students. *SAGE Open, 5*(1). doi: 2158244015573169. Retrieved from http://sgo.sagepub.com/content/5/1/2158

244015573169?utm_content=buffer8b6ba&utm_medium=social&utm_source=plus.google.com&utm_campaign=buffer

McKenney, S., & Reeves, T. C. (2014). Educational design research. In M. Spector, M.D. Merrill, J. Elen, & M.J. Bishop (Eds.), *Handbook of research on educational communications and technology* (pp. 131–140). New York: Springer.

Makoe, M. (2012). Teaching digital natives: Identifying competencies for mobile learning facilitators in distance education. *South African Journal of Higher Education, 26*(1), 91–104.

Nation, P., & Waring, R. (1997). Vocabulary size, text coverage and word lists. *Vocabulary: Description, acquisition and pedagogy, 14*, 6–19.

North, D., Johnston, K., & Ophoff, J. (2014). The use of mobile phones by South African university students. *Issues in Informing Science and Information Technology, 11*, 115–138. Retrieved from http://iisit.org/Vol11/IISITv11p115-138North0469.pdf

Opie, B. (2004). Tertiary education and research in New Zealand. *Minerva, 42*(3), 299–307.

Rezabek, L., & Cochenour, J. (1998). Visual cues in computer-mediated communication: Supplementing text with emoticons. *Journal of Visual Literacy, 18*(2), 201–215. Retrieved from http://www.tandfonline.com/doi/abs/10.1080/23796529.1998.11674539

Schmitt, N., Schmitt, D., & Clapham, C. (2001). Developing and exploring the behaviour of two new versions of the Vocabulary Levels Test. *Language Testing, 18*(1), 55–88.

Shapka, J. D., Domene, J. F., Khan, S., & Yang, L. M. (2016). Online versus in-person interviews with adolescents: An exploration of data equivalence. *Computers in Human Behavior, 58*, 361–367.

Thornton, P., & Houser, C. (2001). Learning on the move: Vocabulary study via email and mobile phone SMS. *World Conference on Educational Multimedia, Hypermedia and Telecommunications, 2001*, (1), 1896–1897.

Tongco, M. D. C. (2007). Purposive sampling as a tool for informant selection. *Ethnobotany Research and Applications, 5*, 147–158.

Turney, L., & Pocknee, C. (2005). Virtual focus groups: New frontiers in research. *International Journal of Qualitative Methods, 4*(2), 32–43.

Wang, F., & Hannafin, M. (2005). Design-based research and technology enhanced learning environments. *Educational Technology Research and Development, 53*(4), 5–23.

Wicks, D., & Sallee, J. (2016). Transactional distance or community of inquiry: A need for a theory of focus in online learning. *Russian-American Education Forum*. Retrieved from http://www.rus-ameeduforum.com/content/en/?task=art&article=1000864&iid=10

Wilkins, D. A. (1972). Grammatical, situational and notional syllabuses. Retrieved from http://files.eric.ed.gov/fulltext/ED136549.pdf

7

THE DEATH OF 'MOBILE LEARNING'

Ian Glover and Helen Rodger

Introduction

Mobile learning (mlearning) as a concept has now come to largely refer to the use of handheld smartphones and tablets to learn in a variety of situations, thereby diverging from earlier research that also considered laptops to be mobile learning devices (Lim, 1999). As mobile and supporting technologies have developed over time, so have the characteristics of mobile learning projects and research, moving from learners accessing and recording simple, static content on basic mobile phones and Personal Digital Assistants (PDAs) through to highly interactive learning experiences using rich media and sophisticated social interaction on highly capable smartphones and tablets. These technical developments have led to mobile learning becoming less content-oriented and instead more interaction-oriented, with egalitarian, informal learning communities based around particular websites or social networks encouraging the development of fully online communities of practice and learning networks (Gikas & Grant, 2013).

This chapter provides a brief summary of the literature related to the shift towards more active pedagogies and the societal and institutional changes brought about by growing access to technology and online networks. In discussing these themes as they relate to mobile learning, we aim to challenge the current emphasis placed on mobile learning as a discrete concept rather than a deeply integrated use of technology, and suggest that this may be holding back the potential for these devices to transform learning more widely. While the authors' views are derived from their experiences within the United Kingdom (UK) higher education (HE) sector, the issues have relevance beyond this setting.

Background Literature

Contemporaneous developments have had a significant influence on the evolution of mobile learning and the use of mobile devices to support learning in HE institutions.

Adoption of Active Pedagogies in Higher Education

A growing trend in HE is the move away from the teacher-driven, didactic, content-delivery model of teaching, towards one that encourages learning by requiring students to be more active participants in the process. Pedagogical theories such as Vygotsky's (1978) Social Constructivism and Siemens' (2005) Connectivism have grown in popularity as the idea of learning being derived from social interaction has become more widespread. Similarly, the use of Authentic Learning approaches, where the learning models 'real-world' contexts, projects and situations as closely as possible (Herrington, 2006), has grown in line with the increased demands for education that is 'relevant' to the professional world and that helps learners develop transferable 'soft skills' in addition to subject knowledge.

These changes to prevailing pedagogies have also been a factor in the development of the concept of hybrid Personal Learning Environments (PLEs) that reflect the interconnections between formal learning and informal, social interaction (Glover & Oliver, 2008). Dabbagh and Kitsantas (2012) suggest that such PLEs can help to foster self-regulated learning in HE learners, in particular. This straddling of formal and informal learning contexts is something for which tablets and smartphones, as personal devices, are ideally suited due to their deep integration into people's lives and ability to facilitate varied interactions and deliver rich content (García-Peñalvo & Conde, 2015).

Mobile learning, with its association of moving between locations and contexts, has also been seen as a powerful tool for realizing these pedagogical aims of personalized, active learning. The technological affordances of mobile devices, in particular their portability and ability to display high-quality media, has made them a commonly used tool in Experiential Learning (Kolb, 1984), ranging from immersive, situated augmented reality (AR) applications (Latif, 2012; Le, Pedro, Lim, Park, Park & Kim, 2015) through to providing a convenient mechanism to record reflections on learning during or soon after an experience of activity (Traxler, 2007). Pedagogical models for mobile learning have been developed by several researchers (Park, 2011; Ozdamli, 2012), while others have developed lenses to enable the classification of mobile learning projects and activities (Kearney, Schuck, Burden, & Aubusson, 2012). Yet, as Lindsay (2016) points out by using Puentedura's (2010) Substitution–Augmentation–Modification–Redefinition (SAMR) model to classify the level of transformation brought to learning activities through the introduction of technology, mobile learning projects are often using technology solely as a more convenient replacement for existing practices rather than making full use of the affordances of the tools to redesign the learning and interaction. While this replacement reflects how we often adopt the use of mobile functionality in our personal lives, that it is largely facilitated by the tutor, and not by the learner, is a limitation of this approach.

Within UK HE specifically, a large part of this move towards active learning can be attributed to a national 'Employability' agenda that emphasizes the development of soft skills and professional behaviours as an expected result of degree programmes (Yorke, 2006). This agenda, driven by both employer requirements and students' employment expectations following graduation, has resulted in increasing alignment of university curricula and learning activities to the prevailing technological tools and ways of thinking and working in relevant professions. Therefore, as workplace adoption of mobile devices increases, so greater integration of the same devices within HE becomes important for supporting Employability initiatives (Burns-Sardone, 2014).

Evolution of Behaviours

Alongside these developments, human behaviours were responding to emerging possibilities afforded by mobile and online technologies (Postman, 1993; Amft & Lukowicz, 2009; Agger, 2011), including the development of a whole new relationship with the World Wide Web. Specifically, there was movement from a publisher–consumer model of information, to one where the publisher is also a consumer and switches freely between these different modes, and where communication is inseparable from consumption (ComScore, 2011). A more egalitarian web evolved, and services appeared that enabled people to very easily share their ideas and views with a worldwide audience. Wikipedia popularized the concept of shared ownership of information, of responsibility, trust and validity by explicitly giving permission to 'change it if it's wrong' (Kittur, Chi, Pendleton, Suh, & Mytkowicz, 2007; Rowley & Johnson, 2013), while blogs made self-publishing highly accessible (Gurak, Antonijevic, Johnson, Ratliff, & Reyman, 2004; Kim, 2005) yet maintained some distinction between content producers and the community of commenters; the rise of social media platforms, such as Facebook and Twitter, where all users have equal 'voice', allowed users to connect and share in highly personalized and flexible ways; and 'instant message' platforms, such as Facebook Messenger and WhatsApp, predominantly used on mobile devices, opened up opportunities for sophisticated, just-in-time organization and communication between individuals, groups and communities (Fuchs, 2012), while facilitating richer relationships (O'Hara, Massimi, Harper, Rubens, & Morris, 2014). Crucially, these capabilities did not wed the user to a place or space, and in some cases exploited the multitasking flexibility of mobile technologies (Oulasvirta, Rattenbury, Ma, & Raita, 2012). These examples are not niche, merely used by few or dabbled with; they are pervasive and largely accessible, and each has made a significant impact on the behaviours of many.

Therefore, learners largely come to HE primed with a personal understanding, context or relationship with each of these examples and, where they make connections and/or see purpose and value, may adapt them to support and enhance their own learning.

This change is already visible in the classroom, as evolving smartphone technologies open up individual routes for personal preference and innovation (Lu, Yao, & Yu, 2005; Jung, 2014). For example: in the pre-smartphone era, the prevalent mode of note-taking in lectures would be to sit, listen, handwrite notes and file those notes away for future reference; post-smartphone adoption, however, other behaviours become possible that encourage autonomous social constructivist and connectivist pedagogies:

- **Sophisticated note-taking** (Schepman, Rodway, Beattie, & Lambert, 2012) tools seamlessly integrate typed text with handwritten notes and rich media, and allow for real-time collaboration. Moving between tasks is an intuitive experience within the tool itself. Notes are available across multiple devices and shared on a cloud for access anywhere, at any time.
- **Synchronous networking** (Young, 2016) is used for collaborative note-taking and online networked discussion, which may be formally directed by the tutor or may be informal, opportunistic and autonomous; it could be limited to the classroom, or stretch beyond into external learning communities, without disrupting the lecture.
- **Critical analysis** can be intuitive, supplementary knowledge building—salient points are researched online in situ (Gevelber, 2016), and the student can reinforce their take on the lecture in real time.
- **Tangible engagement** (Dervan, 2014; Pollard, 2014): Students complete learning activities using their devices, providing the tutor with fast, live data on learning progress; questions and key points can be captured and accessed by the tutor at any point.

It is significant that these behaviours do not need to be mediated by the tutor; the first three may be entirely spontaneous, student owned and student led, and the fourth may be achieved with little management or coordination. This is certainly a shift from the way that mobile learning tends to be more commonly framed.

Additionally, consumer culture has made it acceptable for users to frequently upgrade their mobile hardware, whether to benefit from an advance in new or improved capability that they identify with (Boakye, McGinnis, & Prybutok, 2014), or to experience the hedonistic feeling of a new tool (Venkatesh, Thong, & Xu, 2012). So society has learned that it's acceptable for technology to be disposable, to be fickle with functionality, to experiment, to expect that the thing they want to do but currently can't will soon be possible, and this keeps expectations and behaviours moving forward (Deloitte, 2016).

Discussion

As societies, we have embraced mobile technology in many aspects of our lives, particularly in our social lives, to the point that smartphones and tablets have become our 'window on the world'. This is well illustrated in the phrases 'Pics or it didn't happen' (Silverman, 2016), where the ubiquitous nature of the devices and their capabilities is used implicitly as a counter to online bragging—the implication being that everyone has a smartphone with a camera, so it should be easy to provide evidence for an outlandish claim—and 'There's an app for that' (Chen, 2010), reflecting the vast array of possibilities or tasks that can be achieved via a mobile device. Yet, the methods by which mobile technology has been adopted in HE have differed significantly between educators and learners. Learners have rapidly integrated these tools into informal practice in a way that reflects their instinctual use in other aspects of life, such as by checking an online resource in a face-to-face learning situation or posting a photo of a piece of work to a social network for comment by peers. However, this same level of integration has been much slower to develop in mobile learning initiatives from educators, whose emphasis has mostly been in using the technology as a direct substitute for prior techniques, such as recording lab results on a tablet rather than in a book due to ease of sharing. In fact, research at the authors' institution showed that students prefer to use mobile devices as a means to develop and engage with personal learning networks rather than access university-provided materials and services (Kainz, Powell, & Thorpe, 2011). This highlights the fundamental contradiction with the way mobile learning is viewed by the two primary groups involved in it—for learners it is a personal, instinctual activity embedded in all other learning activity, while for the educators it is a specific type of activity and approach to learning.

In many ways, the growth in ownership of mobile devices and desire to use them as a fully integrated learning tool has been more willingly accepted and embraced by university IT departments than the general academic population. This is particularly evident in UK universities through the widespread development of Bring Your Own Device (BYOD) policies and programmes, where the university provides the backbone infrastructure for people to use their own mobile devices—such as Wi-Fi, power outlets and online access to services and software—rather than locking down services and only allowing access to 'approved' or university-supplied devices. BYOD implicitly acknowledges that people value the affordances provided by mobile devices compared to fixed-location devices, and also want to use their own devices that they have configured to work in the way that they want to work. Further to this, the space between formal and informal learning experiences may now be exploited by learners, either using devices to augment, extend and support their own learning while remaining connected to the formal learning, or as a means for personal organization and creating the 'social glue' (Madge, Meek, Wellens, & Hooley, 2009). In this way mobile use is woven throughout the learning experience.

The personal nature of mobile devices also offers challenges and benefits for their use as learning tools. The widespread adoption of the 'Cloud' has shown that technology is a living, breathing ecosystem that we can access any way we like, for whatever purpose we need, using the tools we have to hand, whether they be mobile, traditional PC or laptop or even a smartwatch. Mobile is merely the chosen window we increasingly look and operate through. The wealth of, frequently free, apps and online services that are available to support particular types of interaction or methods of working means that people are empowered to meet their own objectives using the methods and platforms with which they themselves are most comfortable, rather than needing to use tools selected by someone else (Woodcock, Middleton, & Nortcliffe, 2012). This enables individuals to select the appropriate tools to design their own methods of consuming and sharing information, potentially even having completely different ones for different contexts, such as only using a particular app for social activity while using a different one to achieve the same outcomes in a learning or professional situation. This results in a tension between the ways in which people are using mobile tools and online services to create, manage and develop their own informal learning communities and processes, and the ways in which mobile learning projects in HE often seek to define and control the types of interaction taking place.

The use of mobile devices primarily as a direct replacement of previous practices and methods has been strongly in evidence in formal projects since the inception of mobile learning and suggests that in its current form it may not be the catalyst for disruptive, yet sustainable pedagogical change (Cochrane, 2012). While mobile devices clearly offer increasing affordances that make them valuable and engaging tools to support and encourage learning, the development of specific pedagogies for mobile learning is a case of allowing the technology to define the learning, rather than the alternative and more considered approach where the choice of tool and learning activities is based upon required learning outcomes and the most appropriate method by which they can be achieved (Glover, Hepplestone, Parkin, Rodger, & Irwin, 2016). Combined with the speed at which mobile technology and its use in society is evolving, this relative lack of transformative mobile learning practices suggests that, while research into specific uses of mobile technology is important, there is likely to be more benefit from research into fully integrating mobile learning into existing practices rather than treating it as a specialist area in itself.

However, while a deeper integration of mobile technology into formal learning activities is likely to be of greater benefit to learners and result in more transformative changes to learning and teaching, it is also likely to receive greater resistance from educators. Partly this is a result of the need to cede 'control' to learners over their own learning process rather than managing it directly and the resulting requirement to use pedagogies that focus on learning outcomes without dictating the method by which they are achieved, essentially requiring the educator to be a learning facilitator rather than an information transmitter, something that can be a challenge for HE educators in particular. These are significant issues that may take some time be resolved due to the profound cultural and professional identity changes that they require; however, much as with the key areas of development outlined above, there is also a technological element that needs to be addressed. Principally, the issue is one of developing educators' confidence and capability in integrating mobile technology so that they are able to adapt their teaching to new technological developments, the resultant changes to the ways in which students seek to learn, and the challenges that arise. Ultimately, educators will need to, at a minimum, accept students' use of these technologies and the changes they bring, as the increasing reliance on mobile tools demonstrates that it is not going to be possible to restrict their use by learners in the future.

Conclusion

Technological changes to both mobile devices and online platforms have resulted in social changes that mean that smartphones and tablets have become indispensable devices for many people, yet, while they have become near ubiquitous, their use in HE is still often seen as specialist and distinct from other aspects of teaching and learning. This results in a tension between the way in which learners actually work, and want to work, and how educators want them to work, with learners organically creating their own learning networks and experiences while educators seek to use the tools to create discrete, defined mobile learning experiences. Ultimately, this conflict can only truly be addressed by educators accepting the use of mobile devices to learn however, whenever, wherever and with whomever suits the aims of individual students, as any attempt to actively prevent learners from using these tools is doomed to failure. As the prevailing pedagogical methods in HE move further away from the passive consumption of content delivered simultaneously to groups of learners towards the active construction of new knowledge based on shared investigation and social interaction, mobile learning needs to become a way to support these changes rather than continue to be viewed as a separate pedagogy. It is only by the tools largely 'fading into the background' that the full benefits of mobile devices within HE can be realized.

Implications for Future Practice/Research

By considering mobile learning as a distinct activity and pedagogical approach, there is a danger that educators will fail to see that the devices are rapidly becoming just another part of the overall learning landscape. Just as the web development community has adopted the concept of 'Mobile First'—the understanding that most access to a web page will be through a mobile device and therefore the web page should be designed initially to work well with such devices—so should HE begin to embrace the use of mobile devices as the preference of growing numbers of students and seek to ensure that these tools can be readily integrated into learning and teaching. While a role remains for projects that focus on the use of specific mobile technologies for specific learning outcomes, it is now time for HE to shift the core of its attention toward normalization and accepted practice, toward the development of transferable sustainable skills and behaviours that reflect a culture of lifelong learning.

References

Agger, B. (2011). iTime: Labor and life in a smartphone era. *Time & Society*, *20*(1), 119–136.

Amft, O.D., & Lukowicz, P. (2009). From backpacks to smartphones: Past, present and future of wearable computers. *IEEE Pervasive Computing*, *8*(3), 8–13.

Boakye, K.G., McGinnis, T., & Prybutok, V.R. (2014). Q-TAM: a quality technology acceptance model for technology operations managers. *Operations Management Research*, 7(1–2), 13–23.

Burns-Sardone, N. (2014). Making the case for BYOD instruction in teacher education. *Issues in Informing Science and Information Technology*, *11*(1), 192–200.

Chen, B.X. (2010, October 11). Apple registers trademark for 'There's an app for that'. *Wired*. Retrieved from https://www.wired.com/2010/10/app-for-that/

Cochrane, T. (2012). Secrets of mlearning failures: confronting reality. *Research in Learning Technology*, *20*(Sup1), (123–134).

ComScore. (2011). *Digital omnivores: How tablets, smartphones and connected devices are changing U.S. digital media consumption habits*. Retrieved from http://www.comscore.com/content/download/10633/180227/file/Digital_Omnivores.pdf

Dabbagh, N., & Kitsantas, A. (2012). Personal learning environments, social media, and self-regulated learning: A natural formula for connecting formal and informal learning. *The Internet and Higher Education*, *15*(1), 3–8.

Deloitte. (2016). *There's no place like phone: Consumer usage patterns in the era of peak smartphone*. Retrieved from http://www.deloitte.co.uk/mobileUK/assets/pdf/Deloitte-Mobile-Consumer-2016-There-is-no-place-like-phone.pdf

Dervan, P. (2014). Increasing in-class student engagement using Socrative (an online Student Response System). *AISHE-J: The All Ireland Journal of Teaching and Learning in Higher Education*, *6*(3), 1801–1813.

Fuchs, C. (2012). Social media, riots, and revolutions. *Capital & Class*, *36*(3), 383–391.

García-Peñalvo, F.J., & Conde, M.Á. (2015). The impact of a mobile personal learning environment in different educational contexts. *Universal Access in the Information Society*, *14*(3). 375–387.

Gevelber, L. (2016). How mobile has changed how people get things done: New consumer behavior data. Retrieved from https://www.thinkwithgoogle.com/articles/mobile-search-consumer-behavior-data.html

Gikas, J., & Grant, M.M. (2013). Mobile computing devices in higher education: Student perspectives on learning with cellphones, smartphones & social media. *The Internet and Higher Education*, *19*, 18–26.

Glover, I., Hepplestone, S., Parkin, H.J., Rodger, H., & Irwin, B. (2016). Pedagogy first: Realising technology enhanced learning by focusing on teaching practice. *British Journal of Educational Technology*, *47*(5). 993–1002.

Glover, I., & Oliver, A. (2008). Hybridisation of social networking and learning environments. *Proceedings World Conference on Educational Multimedia, Hypermedia and Telecommunications*. 4951–4958.

Gurak, L., Antonijevic, S., Johnson, L., Ratliff, C., & Reyman, J. (2004). Introduction: Weblogs, rhetoric, community, and culture. *Into the blogosphere*. Minneapolis: University of Minnesota. Retrieved from http://conservancy.umn.edu/handle/11299/172840

Herrington, J. (2006). Authentic e-learning in higher education: Design principles for authentic learning environments and tasks. *Proceedings World Conference on E-Learning in Corporate, Government, Healthcare, and Higher Education (ELEARN)*, 3164–3173.

Jung, Y. (2014). What a smartphone is to me: understanding user values in using smartphones. *Information Systems Journal, 24*(4), 299–321.

Kainz, C., Powell, A., & Thorpe, L. (2011). *Mobile strategy strategic review*. Unpublished internal report. Sheffield Hallam University.

Kearney, M., Schuck, S., Burden, K., & Aubusson, P. (2012). Viewing mobile learning from a pedagogical perspective. *Research in Learning Technology*, 20.

Kim, H. (2005). Blogs as the new media on the internet. *Review of Communication, 5*(2–3), 100–108.

Kittur, A., Chi, E., Pendleton, B.A., Suh, B., & Mytkowicz, T. (2007). Power of the few vs. wisdom of the crowd: Wikipedia and the rise of the bourgeoisie. *World wide web, 1*(2), 1–9.

Kolb, D. (1984). *Experiential learning: experience as the source of learning and development*. Englewood Cliffs, NJ: Prentice Hall.

Latif, F. (2012). CARE: Creating augmented reality in education. *Proceedings EdMedia: World Conference on Educational Media and Technology 2012* (pp. 666–669).

Le, Q.T., Pedro, A., Lim, C.R., Park, H.T., Park, C.S., & Kim, H.K. (2015). A framework for using mobile based virtual reality and augmented reality for experiential construction safety education. *International Journal of Engineering Education, 31*(3), 713–725.

Lim, D. (1999). Ubiquitous mobile computing: UMC's model and success. *Journal of Educational Technology & Society, 2*(4), 125–129.

Lindsay, L. (2016). Transformation of teacher practice using mobile technology with one-to-one classes: M-learning pedagogical approaches. *British Journal of Educational Technology, 47*(5). 883–892.

Lu, J., Yao, J.E., & Yu, C.S., (2005). Personal innovativeness, social influences and adoption of wireless Internet services via mobile technology. *The Journal of Strategic Information Systems, 14*(3), 245–268.

Madge, C., Meek, J., Wellens, J., & Hooley, T. (2009). Facebook, social integration and informal learning at university: 'It is more for socialising and talking to friends about work than for actually doing work'. *Learning, Media and Technology, 34*(2), 141–155.

O'Hara, K.P., Massimi, M., Harper, R., Rubens, S., & Morris, J. (2014). Everyday dwelling with WhatsApp. *Proceedings of the 17th ACM conference on computer supported cooperative work & social computing*, 1131–1143.

Oulasvirta, A., Rattenbury, T., Ma, L., & Raita, E. (2012). Habits make smartphone use more pervasive. *Personal and Ubiquitous Computing, 16*(1), 105–114.

Ozdamli, F. (2012). Pedagogical framework of m-learning. *Procedia – Social and Behavioral Sciences, 31*, 927–931.

Park, Y. (2011). A pedagogical framework for mobile learning: Categorizing educational applications of mobile technologies into four types. *The International Review of Research in Open and Distributed Learning, 12*(2).

Pollard, E.A. (2014). Tweeting on the backchannel of a jumbo-sized lecture hall: Maximizing collective learning in a world history survey. *The History Teacher, 47*(3), 329–353.

Postman, N. (1993). *Technopoly: The surrender of culture to technology*. New York, NY: Vintage Books

Puentedura, R.R. (2010). *SAMR and TPCK: Intro to advanced practice*. Retrieved from http://hippasus.com/resources/sweden2010/SAMR_TPCK_IntroToAdvancedPractice.pdf

Rowley, J., & Johnson, F. (2013). Understanding trust formation in digital information sources: The case of Wikipedia. *Journal of Information Science*, *39*(4), 494–508.

Schepman, A., Rodway, P., Beattie, C., & Lambert, J. (2012). An observational study of undergraduate students' adoption of (mobile) note-taking software. *Computers in Human Behavior*, *28*(2), 308–317.

Siemens, G. (2005, August 10). *Connectivism: Learning as network-creation*. Retrieved from http://www.elearnspace.org/Articles/networks.htm

Silverman, J. (2016, August 11). 'Pics or it didn't happen' – the mantra of the Instagram era. *The Guardian*. Retrieved from https://www.theguardian.com/news/2015/feb/26/pics-or-it-didnt-happen-mantra-instagram-era-facebook-twitter

Traxler, J. (2007). Defining, discussing and evaluating mobile learning: The moving finger writes and having writ. *The International Review of Research in Open and Distributed Learning*, *8*(2).

Venkatesh, V., Thong, J.Y., & Xu, X. (2012). Consumer acceptance and use of information technology: extending the unified theory of acceptance and use of technology. *MIS Quarterly*, *36*(1), 157–178.

Vygotsky, L.S. (1978). *Mind in society: The development of higher psychological processes*. Cambridge, MA: Harvard University Press

Woodcock, B., Middleton, A., & Nortcliffe, A. (2012). Considering the smartphone learner: an investigation into student interest in the use of personal technology to enhance their learning. *Student Engagement and Experience Journal*, *1*(1), 1–15.

Yorke, M. (2006). *Employability in higher education: what it is–what it is not. Learning and employability series*, *1*. York, UK: Higher Education Academy.

Young, J.R. (2016, September 21). How colleges should adapt in a networked age. *The Chronicle of Higher Education*. Retrieved from http://www.chronicle.com/article/How-Colleges-Should-Adapt-in-a/237842

8

ONLINE STUDENTS' PERSPECTIVES OF CONNECTIVITY, NAVIGATION, AND EXPERIENCES WITH MOBILE DEVICES

Billie Jean Holubz

Introduction

The option of taking college classes in an online environment is becoming popular in education, and the demand is increasing due to the increased use of mobile devices throughout many careers (Allen & Seaman, 2014; Chen & Denoyelles, 2013). The proportion of higher education (HE) students taking at least one online course is at an all-time high of 33.5 percent (Allen & Seaman, 2014). People are on the go and becoming dependent upon obtaining knowledge in a mobile format by networking with other people and sources of information (Siemens, 2005). The number of people with cell phones and mobile devices has drastically increased in the last five years. The 21 percent growth of online enrollment far exceeds the 2 percent growth in overall student HE enrollment (Allen & Seaman, 2014). The use of mobile devices in the medical and health-related field are becoming vital to the health-related quality of life for many patients (Steinhubl, Muse, & Topol, 2015). Healthcare professionals are using mobile devices to retrieve medical information and diagnoses, keep records, look up medications, and monitor medical readings from patients (Ventola, 2014). Businesses rely on mobile devices to communicate and network, monitor consumer habits, collect payments, design products, advertise, and track financial records (Donner, 2006; Page, 2013; Paulins, Balina, & Arhipova, 2015; Suprateek & Wells, 2003). The problem is that the need for research about education's role in acclimating students to the efficient use of mobile devices is evident so that students may fully benefit from HE practices and become productive citizens in a technological world (Al-Emran, Elsherif, & Shaalan, 2015; Chen & Denoyelles, 2013; Cheon, Lee, Crooks, & Song, 2012; Cho, Choi, Shin, Yu, Kim, & Kim, 2015; Herrington & Herrington, 2007; Pimmer, Mateescu, & Gröhbiel, 2016).

Literature Review

A review of the literature presented interesting insights that were applicable and confirmed the findings from this research study. "Mobile learning is learning across multiple contexts, through social and content interactions, using personal electronic devices" (Crompton, 2013, p. 4). Paulins, Balina, and Arhipova (2015) stretched the definition of mobile learning further to encompass the concept of learning objects where one is "learning from mobile and learning with mobile" (p. 148). In relation to the purpose and results of this study, the experiences of mobile learning encompassed a learning management system (LMS), which could be interpreted as a learning object as well as a feature of the device.

Research on mobile learning spans many different dimensions, including the mobility of technology, learners, educators, and learning (Al-Emran, Elsherif, & Shaalan, 2015). Confirmation of the need for additional research in mobile learning was extracted from much of the current literature that focused on mobile learning (Al-Emran et al., 2015; Brown, 2006; Cheon et al., 2012; Cho et al., 2015; Herrington & Herrington, 2007; Pimmer et al., 2016; Spada, 2014). Contributions from previous literature coupled with the results of the findings from this study are presented in this chapter.

Theoretical Framework

The motivation of researchers to respond to and build upon the work of other researchers opens new insights into the rapidly changing world of technology. New research generates not only new insights, but also new theories about the behaviors of technology users as well as the daily issues that they face. This researcher considered the Connectivism Theory that views learning within the person, rather than through the basic elements of behaviorism, cognitivism, and constructivism (Siemens, 2005). With a focus on learning within the person, the Navigation Theory was applicable in that educators become mentors and coaches that enabled the student to learn through many different means, locations, and connections of knowledge, rather than learning specific pre-selected information (Brown, 2006). Evaluating the level of technology usage performance, White and LeCornu (2013) proposed a model of a continuum of "visitors" and "residents" that accounts for the ways in which people use technology based on their motivation and context. The model negates previous considerations of "digital natives versus digital immigrants" by using a continuum that places the users on a sliding scale indicated by their habits and experiences of use of and comfort level with technology, rather than in a box based on age and environment.

Connectivism Theory was generated as an alternative to the basic elements of behaviorism, cognitivism, and constructivism that addressed learning within the person. With the emergence of technology and the provisions of learning taking place outside of people, the focus must be placed on the value of what is being learned, beyond the physical presence of the individual. Learning is now dependent upon what may happen in another location that affects the understanding of learners in a different location. Beginning with the individual and building networks between individuals in different locations is a key to optimum learning, understanding, progress, and achievement (Siemens, 2005). Considering past and present paradigm shifts in education, Brown generated a Navigation Theory in which a focus on learning should shift to a new paradigm of navigation, where educators become coaches and mentors to "enable learners to find, identify, manipulate and evaluate information and knowledge to integrate this knowledge into their world of work and life, to solve problems and to communicate this knowledge to others" (Brown, 2006, p. 108). Based upon the theories of connectivism (Siemens, 2005) and navigationism (Brown, 2006), this study explored the mobile connectivity practices and experiences of undergraduate and graduate students enrolled and navigating in an online learning degree program. White and LeCornu's (2013) proposed continuum of the behavior of "visitors" and "residents" based on motivation also played a role in the design and results of this study.

Attitudes and Knowledge

Several important ideas from recent research about attitudes and knowledge were worth noting to set the stage for this study. Since the development of mobile devices as well as improved internet accessibility, mobile learning in HE is increasing. The concept of mlearning presents options for students to complete their education from any location at any time (Al-Emran et al., 2015). Alhassan (2016) reported that findings in a study on ubiquitous learning indicated that students had "highly positive attitudes toward mobile learning, and that they had the necessary technical knowledge to implement mobile learning ... but very little experience in electronic and mobile learning" (p. 176).

Almaiah, Jalil, and Man (2016) quantitatively explored students' perspectives in regards to factors that achieve a high quality of mobile learning systems. They concluded that the following factors determine high-quality mobile learning systems: "system quality (functionality, accessibility, interactivity, interface design, and ease of use), information quality (content usefulness and content adequacy), and service quality (availability, personalization, trust, and responsiveness)" (pp. 5–6). Broadbent and Poon (2015) proposed a systematic review of the self-regulated learning strategies that contribute to the success of mobile learning that can be viewed in conjunction with the factors of Almaiah et al. (2016) that achieve high-quality mobile learning systems. Broadbent and Poon (2015) identified the following self-regulated learning (SRL) strategies in view of online academic achievement: "metacognition, time management, effort regulation, peer learning, elaboration, rehearsal, organization, critical thinking" (pp. 4–5).

Connectivity, Navigation, Experiences

Significant trends in learning that are important when considering the connectivity, navigation, and experiences of participants of mobile learning include the fact that many learners change their major fields of study over time, learning tends to occur more often in informal rather than formal settings, learning is continual and lasts a lifetime, and technology is causing our brains to think in different ways (Siemens, 2005, p. 1). Research relating to connectivity, navigation, and experiences in mobile learning included a study by Gikas and Grant (2013) that indicated that constant connectivity played an important role in the success of using mobile devices for student learning. Even though the study was not exclusive to online education, many similarities to exclusive online education studies emerged that were related to the advantages of using mobile devices for student learning both in a brick-and-mortar setting as well as exclusive online education. Themes of the advantages included "accessing information quickly, constant connectivity, variety in ways to learn, and situated learning" (pp. 21–22). The authors also presented frustrations from learning with mobile devices that included "anti-technology instructors [teachers], device challenges, and devices as a distraction" (p. 23), but found that two-thirds of the students in their study expressed that mobile learning was essential to their academic achievement. However, in another study, results indicated that while students had positive attitudes toward mobile learning, they had very little experience in electronic and mobile learning (Gikas & Grant, 2013). The findings of Al-Emran et al. (2015) indicated that a key component of mobile learning is the ability to learn in anytime, anywhere settings.

Method

The qualitative phenomenological study was implemented in a private eastern United States university that is a leader in online education. The following research questions guided the study: What situations warrant remote accessibility to learning platforms, course content, and communication (with others) in an online learning environment? What are the preferred devices and features that students choose to use for optimum connectivity and navigation in an online learning environment? What methods of connectivity, navigation, and interaction are commonly used by students in an online learning environment?

The empirical phenomenological approach allowed the researcher to obtain "comprehensive descriptions that provided the basis for a reflective structural analysis that portrays the essence of the experience" (Moustakas, 1994, p. 13). After receiving approval from the university's Institutional Review Board to conduct the study, the researcher sent an invitation email describing the details of the study to undergraduate and graduate students who were enrolled in an online education program.

Twenty-six participants from an online learning program at a private eastern United States university and with the minimum age of 18 were accepted for participation. Selection was based upon their order of reply to the recruiting email and willingness to participate in the study. Due to participants' obligations and time restraints, seven of the 26 participants completed their participation in the study. Data was collected for four weeks through interviews and blogs until saturation was reached. The data analysis consisted of sifting through transcripts from interviews and blogs and coding the text to identify recurring patterns and establish overarching themes. From the main themes, the researcher identified sub-themes until an accurate portrayal of the experiences was evident (Miles, Huberman, & Saldana, 2014).

Findings

The findings of this study were consistent from each of the participants who represented undergraduate (one-third), graduate (one-third), and postgraduate degree programs (one-third). Each of the participants had experienced learning previously in a brick-and-mortar higher education setting, but most of the participants had experienced online methods of learning only at the higher education level. This section will address the findings in relation to each of the research questions.

What situations warrant remote accessibility to learning platforms, course content, and communication (with others) in an online learning environment?

The main themes that emerged from the data in regards to situations that warrant remote accessibility to learning platforms, course content, and communication in an online learning environment included work-related responsibilities, family emergencies, weather issues (natural disasters, hurricanes, and snowstorms), power outages, and traveling. Every participant related numerous instances that confirmed a satisfaction with the mobile learning options.

Wendy, a 47-year-old female graduate-level online student, who had received her undergraduate degree in a brick-and-mortar classroom in her home country of Romania, was a full time Special Education teacher enjoying the mobile learning experiences. She stated that she was surprised how well online learning works. She was "impressed with the [university's] modern technology … and support" throughout the journey (personal communication, October 17, 2016). She especially enjoys the "flexibility and freedom to choose hours and when to read" (personal communication, October 17, 2016). The flexibility allowed her to continue her studies without interruption during a major hurricane in her home town. She traveled 130 miles with her mobile devices and never missed an assignment.

Fifty-five-year-old Margaret, who taught at the college level and frequently traveled, "loves being able to take studies wherever I go" (personal communication, October 19, 2016). She emphasized that people who would not be able to go to school can use Skype and WebEx and take online education. She also mentioned a stark contrast to the advantages of online education, explaining the issues that arise with devices in her brick-and-mortar classroom where students are required to participate in class, rather than online. She stated, "Young people have an addiction to technology and people don't put down their devices. They need to find a balance because some of the digital natives never rode a bike or threw a ball. It's sad." Relating to her experiences with mobile learning, she stated, "I love being able to take studies wherever I go. I would love to be an online professor [educator]" (personal communication, October 19, 2016 and online blog posting October 27, 2016).

John, a postgraduate student who formerly served in the military was an experienced online student who had earned his master's degree in 1989 via military satellite. He was impressed with the high standards of this university and "more sophisticated technology that is easy to access when I need it" (personal communication, October 18, 2016). He went on to describe that this online experience allowed him to manage his time so that he could work on homework any time during the day or night and access information in real time.

Reasons such as living in a remote location and experiencing a weak internet signal forced Misty, a undergrad student in her last year of studies, to access course materials from her work computer. The mobility of the online platform allowed her to download/upload assignments and stream videos and presentations from her work location. Attending brick-and-mortar classes would pose a problem to Misty because she must continue to work as a logistics data analyst to earn a living and finance her education. She is grateful for the opportunity to complete her education online because, without the online option, she would not have the privilege of furthering her education.

The illness of his mother necessitated the need for Morey, a student in his last year of college, to consider mobile learning to complete his dream of a college degree. He described the relaxed "free learning" environment that he experiences from mobile learning. You can be "relaxed, yourself, free to get up, get a drink, eat" whenever you desire (personal communication, October 13, 2016).

Joni found mobile learning "a necessity," "huge," and "really valuable and [she] would not have been able to go to school" otherwise. Her love of travel and frequent trips had never been interrupted because of the flexibility of mobile learning. She considers herself "self-motivated" and feels fully comfortable working from a laptop during trips as well as spectating at her kid's sporting events (personal communication, October 23, 2016). "Quality post-secondary education should be available to everyone and mobile learning will allow that to happen by increasing accessibility and flexibility while decreasing cost" (online blog posting on October 29, 2016).

The responses from Mary supported the previous themes and confirmed saturation of the data. Her family situation and work obligations were reasons that warranted the online learning option. She indicated, however, that she finds that the mobile learning environment is "enlightening, but with intense rigor and workload" (personal communication, October 24, 2016).

What are the preferred devices and features that students choose to use for optimum connectivity and navigation in an online learning environment?

The overriding theme from the data relating to this research question was laptops. Every participant owned and used at least two other devices (cell phones and tablets), but the laptop was the primary device that was used for mobile learning. Half of the participants used iOS devices, and the other half used Android/PC devices. Several of the iOS participants mentioned connectivity issues with the LMS that the university uses, but learned that issues were resolved by changing to a different browser. Students' reasons for using a device other than their laptop included (1) checking a grade, and (2) using two devices to complete an assignment—one for research and the other one for typing the assignment.

When asked about features that may be considered when purchasing a device for mobile learning, participants generated the following responses (themes):

- Blackboard (LMS)
- Processing speed
- Visual clarity (large screen)
- Keyboard
- Battery rating
- Video camera
- Windows 10 (Powerpoint, Word, Media Player)
- Touch screen
- Sound system
- Streaming videos and presentations
- Ease of use and adaptability
- Access to iCloud
- Skype
- Recorder
- Outlook and Calendar

It was interesting to note that many of the responses about features also included discussions about the features of the LMS (Blackboard) that was used by the university. None of the participants exclusively discussed the features of their devices without commenting on the ease of the use of the Blackboard LMS and the organized manner by which the curriculum was presented.

What methods of connectivity, navigation, and interaction are commonly used by students in an online learning environment?

Themes related to the methods of connectivity, navigation, and interaction were numerous and covered many perspectives related to mobile learning. Participants used the terms access, internet, Wi-Fi, travel, locations, compatibility, Help Desk, communication, support, responsibility, and collaboration consistently throughout the interviews. In regards to connectivity, they indicated that access to their coursework was 24/7 from any location in which they had internet access. The convenience of taking course work with them wherever they went and having the option to complete course work at any time of the day and night were key factors that attracted them to mobile learning. Participants indicated that communication and collaboration with educators and peers were less than they would have been if they were attending school in a brick-and-mortar classroom. On the other hand, several of the participants felt secure and comfortable in communicating through Discussion Board Forums rather than orally in a classroom:

> I'm not a good test taker. The Discussion Board Forum gave me the opportunity to show that I know the material.
>
> *Personal communication, October 16, 2016*

> The Discussion Board Forum is a powerful tool that worked well to [allow me] to express myself openly and freely in writing without feeling judgement. It's powerful.
>
> *Personal communication, October 19, 2016*

> I believe the idea of mobile learning will be something that more learning institutions will offer. Likewise, I think more students will pursue advanced degrees because of the convenience of this learning venue. However, while information access and advancement will evolve, the connections made through human contact will dissipate to some extent.
>
> *Online blog posting, October 26, 2016*

Conclusion

The findings from this study provided interesting and valuable information in regards to the mobile learning program of a leading online university. Based on the theories of connectivism (Siemens, 2005) and navigationism (Brown, 2006), this study explored the mobile connectivity practices and experiences of university students enrolled in and navigating an online learning degree program. The participants indicated that by using many tools and resources, they were successful in connecting and navigating throughout the online courses and requirements. Repeatedly discussed by the participants was the fact that the LMS was "organized and easily accessible" to gather the required resources for completing course work. If difficulties arose, a Help Desk (online technical assistance), tutorial, library resources, and educators offered support to students. These results supported Brown's Navigation Theory (2006) with the concept of educators becoming mentors and coaches who enable students to learn through many different means, locations, and connections of knowledge. From a plethora of locations and situations that warranted travel, participants continued learning regardless of the proximity to a brick-and-mortar classroom. Regardless of the learners' locations, course materials and a level of support from mentors and educators were available 24 hours a day.

Participants' ages ranged from early 20s to late 50s, and the results of the data indicated that participants' comfort level with using technology supported the concept of White and LeCornu's (2013) proposed continuum of the behavior of "visitors" and "residents" based on motivation. Each of the participants, regardless of age, considered himself/herself "extremely comfortable" with the use of technology. The participants reported high motivation levels with respect to their uses of mobile devices in learning. The participants' focus on completing online coursework superseded any inadequacies or fears that they may have faced during their online learning experiences. During several of the interviews, participants revealed that they were better adapted than they had realized to using technology and were using technology for many everyday tasks such as email, calendars, shopping lists, note-taking, reading, banking, social media, instant messaging and texting, etc. Participants indicated that a lack of confidence or inexperience did not hinder the extent to which they utilized devices in regards to coursework. Utilizing a mobile device to complete coursework became second nature to them, which indicated that they would fall towards the "residents" side of the White and LeCornu (2013) continuum. Each of the participants of the study utilized their laptop(s) most frequently for mobile learning, which coincided with the work of Herrington and Herrington (2007). The use of phones and tablets was secondary—for purposes such as checking grades, announcements, or postings. Only one participant used a desktop less than 25 percent of the time that she worked on school work. "On the go," "mobile," "taking learning anywhere," and "faster access" were several of the descriptions of learning with mobile devices.

The results of this qualitative study also supported the quantitative findings of Almaiah, Jalil, and Man (2016) in regards to the factors that achieve high-quality mobile learning systems. The participants' frequent references to and descriptions of the features of Blackboard presented a positive image of the high quality of education that the participants received. Many of the themes that emerged from this study (accessibility, ease of use, availability, support) matched the themes from the works of Almaiah, Jalil, and Man (2016). Reference to the accessibility, ease of use, and availability of the LMS were recurring themes in each of the interviews.

The participants' frequent mention of SRL and related terms supported the SRL of Broadbent and Poon (2015). Numerous participants emphasized the importance of time management, self-motivation, organization, rigor, and critical thinking as keys to the success of their online learning experiences.

This study on mobile learning spanned many different dimensions including the mobility of technology, learners, educators, and learning by examining university students' perspectives of navigation, connectivity, and experiences with mobile learning (Al-Emran, Elsherif, & Shaalan, 2015). In agreement with mobile learning researchers, additional research in mobile learning is needed (Al-Emran, et al., 2015; Brown, 2006; Cheon et al., 2012; Cho et al., 2015; Herrington & Herrington, 2007; Pimmer et al., 2016; Spada, 2014).

Implications

Several significant key points that may be relevant to higher education stakeholders resulted from this study.

- The human responsibilities and demands of life indicate a need for mobile learning to keep pace with educational needs and employment opportunities.
- Due to increased responsibilities and time restraints in the personal lives of students of higher education, technology has become a major priority in learning.
- Well-organized LMSs are vital to the success of mobile learning higher education programs.
- Higher education students benefit not only from well-organized LMSs, but also from substantial support teams that include 24-hour technical support and well-equipped, accessible educators.
- Students of mobile learning adapt to the features of mobile devices as needs arise.

Recommendations for Future Research

Recommendations for future studies include research on various types and features of LMSs because of the recurring theme of Blackboard (LMS) throughout this study. Further research that compares and contrasts the features of the LMSs may contribute richly to university administrators as they design online learning environments. Finally, research in teacher education training programs involving mobile learning are vital to implementing, assessing, and improving mobile learning initiatives.

References

Al-Emran, M., Elsherif, H. M., & Shaalan, K. (2015). Investigating attitudes towards the use of mobile learning in higher education. *Computers in Human Behavior*, *56*, 93–102. doi:10.1016/j.chb.2015.11.033

Alhassan, R. (2016). Mobile learning as a method of ubiquitous learning: Students' attitudes, readiness, and possible barriers to implementation in higher education. *Journal of Education and Learning JEL*, *5*(1), 176. doi:10.5539/jel.v5n1p176

Allen, I. E., & Seaman, J. (2014). *Grade change: Tracking online education in the United States.* Babson Survey Research Group and Quahog Research Group, L.L.C.

Almaiah, M. A., Jalil, M. A., & Man, M. (2016). Empirical investigation to explore factors that achieve high quality of mobile learning system based on students' perspectives. *Engineering Science and Technology, an International Journal, 3*(4). doi:10.1016/j.jestch.2016.03.004

Broadbent, J., & Poon, W. (2015). Self-regulated learning strategies & academic achievement in online higher education learning environments: A systematic review. *The Internet and Higher Education, 27*, 1–13. doi:10.1016/j.iheduc.2015.04.007

Brown, T. H., (2006). Beyond constructivism: navigationism in the knowledge era. *On the Horizon, 14* (3), 108–120. doi.10.1108/10748120610690681

Chen, B., & Denoyelles, A. (2013). Exploring students' mobile learning practices in higher education. *Educause Review*. Retrieved from http://er.educause.edu/articles/2013/10/exploring-students-mobile-learning-practices-in-higher-education

Cheon, J., Lee, S., Crooks, S. M., & Song, J. (2012). An investigation of mobile learning readiness in higher education based on the theory of planned behavior. *Computers & Education, 59*(3), 1054–1064. doi:10.1016/j.compedu.2012.04.015

Cho, Y. H., Choi, H., Shin, J., Yu, H. C., Kim, Y. K., & Kim, J. Y. (2015). Review of research on online learning environments in higher education. *Science Direct, 191*(2015). doi:10.1016/j.sbspro.2015.04.634

Crompton, H. (2013). A historical overview of mobile learning: Toward learner-centered education. In Z. L. Berge & L. Y. Muilenburg (Eds.), *Handbook of mobile learning* (pp. 3–14). Florency, KY: Routledge.

Donner, J. (2006). The use of mobile phones by microentrepreneurs in Kigali, Rwanda: Changes to social and business networks. *Information Technologies & International Development 3*(2), 3–19.

Gikas, J., & Grant, M. M. (2013). Mobile computing devices in higher education: Student perspectives on learning with cellphones, smartphones & social media. *The Internet and Higher Education, 19*, 18–26. doi:10.1016/j.iheduc.2013.06.002

Herrington, A., & Herrington, J. (2007). Authentic mobile learning in higher education. *AARE 2007 International Education Research Conference, 28 November 2007*, Fremantle, Western Australia. Retrieved from http://researchrepository.murdoch.edu.au/5413

Miles, M. B., Huberman, A. M., & Saldana, J. (2014). *Qualitative data analysis.* Los Angeles: Sage.

Moustakas, C. (1994). *Phenomenological research methods.* Thousand Oaks, CA: Sage.

Page, T. (2013). Use of mobile device apps in product design. *International Journal of Green Computing, 4*(1), 18–34. doi:10.4018/jgc.2013010102

Paulins, N., Balina, S., & Arhipova, I. (2015). Learning content development methodology for mobile devices. *Procedia Computer Science, 43*, 147–153. doi:10.1016/j.procs.2014.12.020

Pimmer, C., Mateescu, M., & Gröhbiel, U. (2016). Mobile and ubiquitous learning in higher education settings. A systematic review of empirical studies. *Computers in Human Behavior, 63*, 490–501. doi:10.1016/j.chb.2016.05.057

Siemens, G. (2005). Connectivism: A learning theory for the digital age. *International Journal of Instructional Technology and Distance Learning, 2*(1), 3–10.

Spada, K. (2014, May 20). Higher education in the digital age. (Review of the book *Higher education in the digital age* by W. G. Bowen, Ed.). doi:10.1109/TPC.2014.2311873.

Steinhubl, S. R., Muse, E. D., & Topol, E. J. (2015). The emerging field of mobile health. *Science Translation Medicine*, 7(283), 283rv3. doi: 10.1126/scitranslmed.aaa3487

Suprateek, S., & Wells, J. D. (2003). Understanding mobile handheld device use and adoption. *Communications of the ACM 46*(12), 35–40. doi:10.1145/953460.953484

Ventola, C. L. (2014). Mobile devices and apps for health care professionals: Uses and benefits. *Pharmacy and Therapeutics*, *39*(5), 356–364. Rebecca Johnston, 2009.

White, D. S. & LeCornu, A. (2013). Visitors and residents: A new typology for online engagement. *First Monday*, *16*(9). Retrieved from http://firstmonday.org/ojs/index.php/fm/article/view/3171/3049

9

INTEGRATING MOBILE TECHNOLOGIES FOR LEARNING

Thoughts from Educators Who Followed *The 12 Apps of Christmas* Course @ DIT

Frances M. Boylan

Introduction

The integration of mobile technologies into learning, teaching, research, and assessment activities is referred to as mobile learning, or mlearning, which, for the purposes of this book has been defined as "learning across multiple contexts, through social and content interactions, using personal electronic devices" (Crompton, 2013). While the field of mobile learning is relatively new, quite a lot of research has been undertaken on the topic to date. A search of any peer reviewed journal dedicated to learning and teaching related topics will display an array of articles reporting on research, of varying levels of complexity, that has been carried out around, for example, practitioners' experiences of employing mlearning in the classroom (Falloon, 2015); student perceptions of mlearning (Gikas & Grant, 2013); theories of mlearning (Wang & Shen, 2011); benefits of mlearning in general (Mehdipour & Zerehkafi, 2013) and for specific subjects (Wu et al., 2012); pedagogical frameworks for mlearning (Kearney et al., 2014); evaluations of the integration of specific apps in the classroom or field (Chittaro & Sioni, 2014); and employing mlearning to deliver continuing professional development (CPD) in different sectors (Walsh et al., 2013), to name just a few.

The New Media Consortium (NMC) also carries out extensive research around the use of new media and technologies for learning. One of their projects, the Horizon Project, is an ongoing research project and its yearly higher education report identifies and describes "emerging technologies likely to have an impact on learning, teaching, and creative inquiry in education" (NMC, 2016). A progression in expectations around the integration of mobile learning for learning and teaching has been evident over the last number of reports. In 2010, for example, most of the technologies mentioned as ones to watch were either somewhat or totally dependent on mobile learning and included electronic books and augmented reality. By the 2013, 2014, and 2015 reports, mlearning is a given and the reports begin to focus more on what is possible with mobile learning rather than just the technology itself, so, flipped classrooms and gamification, for example, are mentioned. The 2016 report goes on to mention Bring Your Own Device (BYOD) as well as learning analytics and adaptive learning, which make use of mobile learning and the student data that can be gathered through such online learning environments. It is suggested that these will be adopted increasingly in higher education institutions in one year or less.

I would argue, however, that many educators, and the higher education institutions in which they work, both in Ireland and elsewhere, are a few years away yet from achieving what the Horizon Report expects will be the norm in one year or less, and the data gathered from participants through the course "*The 12 Apps of Christmas*" supports this view. But why is this the case? What is blocking the further development of mobile learning, and what could we start to do about that? This chapter will explain what *The 12 Apps of Christmas* course was all about and detail the nature of the data gathered, before presenting what that data told us around educators' current mlearning practices, and their fears and needs around its integration.

The 12 Apps of Christmas: An Overview

In December 2014, December 2015, and again in December 2016, the Learning, Teaching and Technology Centre (LTTC) at the Dublin Institute of Technology (DIT), Dublin, Ireland, compiled and delivered the course entitled "*The 12 Apps of Christmas*". This course is a continuing professional development (CPD) opportunity designed to raise awareness among educators of the benefits mobile technologies and mobile apps present for learning, teaching, and assessment; to upskill anyone interested in experiencing, exploring, and learning more about integrating mobile learning technologies into their teaching, learning, and assessment practices; and to help connect like-minded individuals and provide them with an opportunity to expand their personal learning networks and start conversations with others interested in this emerging field.

Although participants were required to register for the course each year, it was completely free to follow and was openly accessible online. All content was licensed under the Creative Commons License CC BY-NC-SA 2.0, and when the courses ended the websites remained available for use as open educational resources (see http://www.dit.ie/the12appsofchristmasarchive/ for links).

The course format for each year was the same. Each morning, over 12 consecutive weekdays, a page on the website was released, which contained information about a specific mobile app. The pages were designed in such a way that participants could gather the most basic information about the app-of-the-day in approximately ten minutes but could delve into further details and practice tasks if they so wished.

In its inaugural year (2014), the course was aimed at educators specifically and showcased 12 mobile apps that could be employed to enhance and transform learning activities, in a wide variety of ways, both inside and outside of the classroom. Each day, the sample learning and teaching ideas presented around the app were aligned with the SAMR (**S**ubstitution, **A**ugmentation, **M**odification, **R**edefinition) model created by Dr. Ruben Puentudura, highlighting that app's potential for enhancing, modifying, and/or redefining the learning experience. 700 participants worldwide registered to take part.

The following year, students of university-going age and beyond were the target audience, although support material was provided also for educators and support staff. The theme was inspired by Olsen & ideasLAB's comment (2011): "[W]e need to develop learners who are skilled at personalising their learning, as the changing nature of knowledge means this is a fundamental skill for today's workforce". And so, the participants were introduced to 12 different apps over the course of the 12 weekdays and shown how these apps could be used to help personalize their learning. For educators, ideas were provided on how to create a learning environment within which their students could harness the opportunities mobile apps offer for personalizing learning. Almost 2,000 participants registered to follow the course, 982 of whom were educators. Finally, in 2016, the course was aimed once more at academics and was a collaborative effort this time between DIT and academics from nine institutes of higher education worldwide. Each day participants were presented with a case study, which introduced a mobile app and outlined how the author/s had integrated it into their own teaching, learning and/or assessment practices; noted the benefits for their students; presented student feedback gathered; and offered recommendations for use. 3,078 people registered to follow the course.

The course design each year was influenced by social constructivist learning theory. Opportunities were built into the course to encourage the participants to engage with each other, to collaborate, share, and learn. Various strategies were employed to help build a sense of community, which included sending a welcome email following registration to add a personal touch and motivate participants; sending a short reminder email each morning of the course to encourage them to engage with the content and tasks; using the dedicated Twitter hashtag #12appsDIT; and enabling comments on the WordPress website. Some of the daily optional tasks each year were gamified and in 2015 prizes were offered as well, and this generated interest and also some fun among the participants.

With the participants' permission, and the approval of the research ethics committee at DIT, a wide range of data was collected around the course each year, which included feedback given on registration forms and feedback surveys. This chapter will draw on elements of that combined data, most specifically:

- Answers given to a question in the registration forms that asked why the participants were interested in taking part in this professional development opportunity;

- Responses given to a question in the feedback surveys around whether or not participants felt that some type of training or professional development was essential in order to integrate mobile apps in a pedagogical manner, coupled with related qualitative feedback given in the open comments question;
- Participants' opinions on the barriers to integrating mobile apps received in the 2015 feedback survey and in the 2014 follow-up survey sent almost a year after that year's course had run; and finally,
- Feedback given in the 2014 follow-up survey around whether or not the course had influenced or changed how they would integrate mobile apps in their learning, teaching, and assessment practices.

The feedback from the 2016 iteration of this course had not yet been gathered at the time of writing and so has not been included here. Thematic analysis was undertaken on the aforementioned qualitative data. With regard to the data gathered during the 2015 iteration of the course, analysis was confined to the data provided by those who had identified themselves as educators specifically. Where the same question was posed to each of the two cohorts, the same themes were used during the analysis phase to allow for meaningful comparisons and discussion. Some statistical analysis was undertaken also.

Integrating Mobile Apps, Mobile Literacy, the Value of Training, and Barriers to Mlearning

Each year during the registration process, the participants were asked why they were interested in taking part in this 12-day CPD opportunity, and the analysis of the responses showed that there were three distinct reasons for doing so. The main reason given each year was to use the course as a means through which to learn how to integrate apps into their professional practices so as to improve their teaching and engage their students in the learning process. This reason accounted for the responses from 45 per cent (n = 317) of the 2014 cohort, and 40 per cent (n = 394) of the 2015 cohort. What was really interesting, however, was that the comments made showed quite a diverse range of opinion regarding their self-assessed level of mobile literacy, by which I mean the range of knowledge and skills they felt they possessed that would allow them to engage comfortably with mobile technologies for learning and teaching purposes. For example, one participant commented: "I have never used an app in my life and I really need to get with it. I want to get up to date with the latest teaching innovations and enhance my interaction with the students". Another mentioned, "I need to improve my knowledge of ICT, social media, and other mobile learning techniques to improve my teaching and learning. There is so much out there it is hard to know where to start". A third echoed these sentiments, saying, "I feel I need to update my knowledge of teaching tools. PowerPoint is great but I am sure there are so many apps to make teaching even greater".

Others seemed a little more comfortable with mobile technologies on an individual level, but their comments indicated that they were unsure as to how to make use of that knowledge and skills effectively in the classroom. One said that they had registered for the course because "there is so much on offer in the digital apps world it's important to know as an educator what helps and what is white noise", while another participant was "interested to find new ways to make learning attractive to students and to encourage them to take control of their learning".

At the opposite end of the spectrum, a large number of the educators implied that they were employing at least some (mobile) technologies already in the classroom, and were looking for new and innovative ideas. "I have a very keen interest in educational technologies, technology in general, and I am always looking at new ways to integrate technology into the learning experience for today's students", said one. Another noted, "I would like to be introduced to new and existing apps that may be useful in enhancing the students' learning experience".

A second reason participants gave for registering on *The 12 Apps of Christmas* was to stay up to date, develop skill sets, and learn something new, and these accounted for 27 per cent of the participants in 2014 (n = 187), and 29.5 per cent of the participants in 2015 (n = 290). Essentially, they too were using the course as a form of CPD, just like their peers discussed above. However, what made this group a little different was that in their comments they gave little to no commitment to integrating mobile apps into their learning and teaching practices following the course. They were using the course purely as a means to find out more and to keep abreast of developments from more of an academic point of view, noting that "CPD is essential", that they were "looking to see how the apps are used to improve student learning", and they would like "to know more about their [mobile apps] educational potential".

The third and final reason given for registering to follow the course was simply curiosity. There were those who were curious to find out what the course was all about, and those who wanted to see how this short bite-sized approach to CPD would work in practical terms on a daily basis. Comments made included: "Need to see this innovative idea first hand", "A colleague recommended it to me as a ground-breaking experience and so I wanted to get on board", and "I think it sounds really interesting and educational. I have not seen anything like it before so I am curious now to participate". In 2014, this group of participants accounted for just 8 per cent of the responses (n = 55) and in 2015, 9.6 per cent (n = 94).

As interesting as it is to know *why* educators registered for the course, their responses do not tell us whether or not they think that undertaking training, of any description, around mlearning is essential in order to integrate mobile technologies in a pedagogical manner. So, the feedback survey sent following the course each year asked them this very question. The 2014 cohort responded with a resounding 'no', with 60 per cent of them (n = 109) feeling that training is not essential. Only 38 per cent (n = 69) felt it is essential and 2 per cent (n = 3) chose not to answer. The 2015 cohort disagreed with them, however: 69 per cent (n = 70) felt that training is essential with just 29 per cent (n = 30) suggesting that it is not essential; 2 per cent (n = 2) chose not to answer. Given the huge differences of opinion here on the issue of training, this feedback is not particularly helpful. What was of more value, however, was the feedback participants from the 2014 cohort gave almost 12 months after they had taken part in the course. Asked whether or not *The 12 Apps of Christmas* had influenced them in how they had gone about integrating mobile apps into their teaching practices, the feedback was overwhelmingly positive. This was in stark contrast to the fact that straight after they had completed the course a year earlier the majority of that group had felt that training was not essential in order to integrate mobile technologies in a pedagogical manner. It was as though the benefits of the course had only become apparent over time, once they had started integrating mobile technologies. To quote some of the feedback:

> Yes, based on the SAMR model early adoption was simply substitution, but now those apps have freed up some of my effort to allow me to introduce extra material, concepts, and practice into classes. I now regard them as augmenting my teaching.
>
> Totally it is not something I would have thought about prior to this course.
>
> The way the course was structured really helped to link technology to pedagogy and has made [me] even more aware of the relevance of such linkages.
>
> I found that it helped me discover some apps, explore them, and decide if/how to use them—it was great to have the pedagogical basis linked.
>
> I think the main thing for me was losing app-phobia! I just knew there were millions of them out there and I had no idea what they did. I'm not blazing a trail by any means but I do feel I can have a go now and I can see the potential.

Other research undertaken in the area in recent years has also borne out the fact that training around the integration of mlearning is necessary and valuable. Chen and Denoyelles' (2013) exploration of students' mobile learning practices concluded that even though students and educators alike use mobile devices on a regular basis, both groups need technical, logistical, and pedagogical support adopting such technologies if the effectiveness of mobile learning, as a teaching and learning approach, is to improve. MacCullum et al. (2014) arrived at a similar conclusion when they undertook a study around factors that impact on the adoption of mobile learning, noting that "teaching lecturers to use the mobile technology is also important" and that their study had indicated that "advanced mobile literacy played a significant role on adoption". This notion of the importance of supporting staff in the adoption of new technologies is not new, of course, or specific either to the integration of mobile technologies. Carr (1999) writes that "in addition to a strong stable advocacy needed to ensure the conditions necessary for technology adoption and diffusion, training in its technical aspects and application to real needs is crucial to its integration beyond the innovators and early adopters".

Unfortunately, educators face a range of barriers when integrating mobile technologies into their teaching and assessment practices and these tend to have a negative impact on their efforts. When asked what these barriers are, each cohort from *The 12 Apps of Christmas* made very similar suggestions. The most common barrier experienced by individuals was finding the time in their busy schedules to plan for the effective integration of mobile technologies. It was also felt that the lack of awareness and knowledge among educators (including low levels of mobile literacy) acted as a barrier, as did lack of confidence in selecting valuable and appropriate apps, and the fear of losing student attention to non-academic apps during class. Battery issues, storage problems, not all apps being cross platform, and the availability of devices among the student body were some of the technical barriers noted. However, some of the barriers noted by the participants sat at an institutional level as well, and included stable Wi-Fi or the lack thereof in some institutions, the absence of time allocation to offset time spent working with students online in lieu of a traditional face-to-face session, and the lack of BYOD (Bring Your Own Device) policies that support the widespread integration of mobile technologies for learning. Interestingly, the lack of training was mentioned as a barrier also, lending further support to the argument made earlier around the importance of its provision.

Summary

So what has the exploration of this data told us? Well, it's apparent that there is a lot of genuine interest in the integration of mobile technologies for learning and teaching, and there is a certain understanding that there can be a benefit for student learning when these technologies are integrated effectively. Having said that, mobile literacy appears to be an issue and quite a diverse range of skill levels was reported. Despite the differing opinions among the participants as to whether training is essential or not, the value of training, and the role it can play in the effective integration of mlearning, has been recognized somewhat by them. The barriers at institutional level, however, are a little worrying. As one participant commented, "The university's narrow range of technology that is supported by central services [is a barrier]—so if I adopt the use of apps with students and something goes wrong I can't necessarily get suitable support to help fix problems". It could be argued that the institutional barriers are the genuine roadblocks to further development. Individual educators can overcome, or at least work around, the vast majority of the other barriers noted above, but are quite powerless when faced with barriers at an institutional level, such as the lack of stable Wi-Fi or the absence of a well-supported BYOD policy.

The Irish Context—Where to Now?

The most recent work done in the educational technology space, at a national level, has been by the National Forum for the Enhancement of Teaching and Learning in Higher Education. In 2015, and following a period of sectoral consultation, they launched a report entitled *A Roadmap for Enhancement in a Digital World 2015–2017*, and this report presents action-oriented ways in which the higher education sector in Ireland can build digital capacity and work towards achieving a shared culture of digital learning and digital innovation. It also encourages dialogue within institutions so that the agreed vision, principles, and priorities set out in the report can "be translated into tangible and transformative actions on the ground", and that educators "will be fully enabled to use digital technologies … in order to enhance student learning" (National Forum, 2015). This report focuses completely on guiding the sector for technology integration *in general*, however. We seem, as a nation, to be struggling still at that level and have not really cracked the "technology enhanced learning, teaching, and assessment" nut properly yet, and we will need to do this before we can focus on further enhancements such as mobile learning. While a word search in the report for "mlearning", "BYOD", or "bring your own device" does not yield any results at all, a search for the word "mobile" shows two results. The first appears within a section arguing for the constant development of educators' digital literacies and "mobile devices" are given as an example of the types of innovation that "need to be recognised and assessed appropriately". The second mention comes amid a short section acknowledging the changing environment within which students now live and learn. There it is noted: "[T]he proliferation of mobile devices such as tablets and smartphones is one area that is impacting on the technological infrastructure. This is evidenced in the huge growth in demand for Wi-Fi that is being reported across the sector", before going on to say that the pedagogical opportunities that mobile devices offer are only "beginning to be recognised, explored and exploited". More recently, the National Forum has seconded an individual to work on a national project called "Enabling Policies for Digital Teaching and Learning", which recognizes that the lack of certain policies can present "complex issues and challenges which may affect or inhibit the adoption of digital teaching and learning approaches by staff and institutions" (National Forum, 2016). The project will map the existing policy infrastructure in Ireland for digital teaching and learning at both the sectoral and institutional level; provide a suite of high-level policy frameworks suitable for localization at institutional level; and identify case studies of good practice.

While it would appear that mlearning, at every level, is at a very embryonic stage yet in Ireland, there are still things that higher education institutions here could begin doing now to foster the right conditions under which mlearning will grow over time. For ideas we can look to JISC, a not-for-profit company based in the United Kingdom (UK) that provides useful advice to the UK higher education sector on all matters to do with educational technologies. JISC have put together a useful guide to help institutions embrace mobile learning (JISC, 2015). This guide highlights the stages an institution needs to go through "to incorporate mobile learning into [their] organisation's strategy and put in place a rigorous implementation plan". It notes different pedagogical frameworks for mobile learning, discusses overcoming barriers and finding enablers, provides achievable steps to mobile learning adoption, and highlights technical considerations along with some guiding principles. The guide also suggests providing ongoing training for all stakeholders.

Alongside this guide, however, JISC also highlights the fact that a key challenge for institutions is the development of a coherent BYOD policy and they provide a link to really useful key findings and solutions (JISC, 2016). This is, in my opinion, where Irish institutions really need to start, and hopefully the current work of the National Forum, as referenced earlier, will help with this. For those who do not already have one, the development of a comprehensive BYOD policy specifically, that has, most importantly, a "focus on academic practice" (JISC, 2016), will go a long way towards addressing many of the barriers faced by their educators integrating mlearning, as reported by the participants of *The 12 Apps of Christmas*, allaying their fears and making it possible to design solutions to address their needs and wants. Without that in place, all other efforts by educators and educational technologists alike will just operate in a vacuum, and will not be as effective, or as sustainable into the future, as they should be. The BYOD policy really is the document that holds it all together.

References

Carr Jr., V. H. (1999). Technology adoption and diffusion. *The Learning Centre for Interactive Technology*. Retrieved from http://www.au.af.mil/au/awc/awcgate/innovation/adoptiondiffusion.htm

Chen, B., & Denoyelles, A. (2013). Exploring students' mobile learning practices in higher education. *EDUCAUSE Review Online*. Retrieved from http://bit.ly/2aVviCC

Chittaro, L,. & Sioni, R. (2014). Evaluating mobile apps for breathing training: The effectiveness of visualization. *Computers in Human Behavior*, *40*(5), 56–63.

Crompton, H. (2013). A historical overview of mobile learning: Toward learner-centered education. In Z. L. Berge & L. Y. Muilenburg (Eds.), *Handbook of mobile learning* (pp. 3–14). Florence, KY: Routledge.

Falloon, G. (2015). What's the difference? Learning collaboratively using iPads in conventional classrooms. *Computers & Education*, *84*(C), 62–77.

Gikas, J., & Grant, M. (2013). Mobile computing devices in higher education: Student perspectives on learning with cellphones, smartphones & social media. *The Internet and Higher Education*, *19*, 18–26.

JISC. (2015). Mobile learning. Retrieved from https://www.jisc.ac.uk/guides/engage-students-with-mobile-learning

JISC. (2016). JISC digital student – challenge 5: coherent policies on 'bring your own'. Retrieved from https://digitalstudent.jiscinvolve.org/wp/challenges/challenge-coherent-policies-on-bring-your-own/

Kearney, M., Burden, K., & Rai, T. (2014). Investigating teachers' adoption of signature mobile pedagogies. *Computers & Education, 80*, 48–57.

MacCallum, K., Jeffrey, L., & Kinshuk. (2014). Factors impacting teachers' adoption of mobile learning. *Journal of Information Technology Education: Research, 13*. Retrieved from http://www.jite.org/documents/Vol13/JITEv13ResearchP141-162MacCallum0455.pdf

Mehdipour, Y., & Zerehkafi, H. (2013). Mobile learning for education: Benefits and challenges. *International Journal of Computational Engineering Research, 3*(6), 93–101.

The National Forum for the Enhancement ofTeaching and Learning in Ireland Education. (2015).Teaching and Learning in Irish Higher Education:A Roadmap for Enhancement in a Digital World 2015–2017. Retrieved from http://www.teachingandlearning.ie/wp-content/uploads/2015/03/Digital-Roadmap-web.pdf

The National Forum for the Enhancement ofTeaching and Learning in Ireland Education. (2016). Building digital capacity: Enabling policies for digital teaching and learning. Retrieved from http://www.teachingandlearning.ie/wp-content/uploads/2016/03/PubToREnablingPolicies.pdf

New Media Consortium. (2016). NMC horizon report. *2016 higher education Edition*. Retrieved from http://www.nmc.org/publication/nmc-horizon-report-2016-higher-education-edition/

Olsen, R. & ideasLAB. (2011). Understanding pedagogies for contemporary teaching & learning: an ideasLAB white paper. Retrieved from http://bit.ly/1WN2pdu

Walsh, C., Power, T., Khatoon, M., Biswas, S., Paul, A., Sarkar, B., & Griffiths, M. (2013). The 'trainer in your pocket': mobile phones within a teacher continuing professional development program in Bangladesh. *Professional Development in Education, 39*(2), 186–200.

Wang, M., & Shen, R., (2011). Message design for mobile learning: Learning theories, human cognition and design principles. *British Journal of Educational Technology, 43*(4), 561–575.

Wu, Po-Han., Hwang, Gwo-Jen., Su, Lisng-Hao., & Huang,Yueh-Min. (2012).A context-aware mobile learning system for supporting cognitive apprenticeships in nursing skills training. *Journal of Educational Technology & Society, 15*(1), 223–236.

10

MAINSTREAMING MOBILE TEACHING INNOVATION IN A RESOURCE-CONSTRAINED CONTEXT

Changing Access, Shifting Practice

Cheryl Brown, Genevieve Haupt, and Aditi Hunma

South African Context

Over the past decade, universities in South Africa recognized the role of educational technologies as tools to facilitate teaching and learning (Czerniewicz & Brown, 2009; Bozalek et al., 2013). While the integration of personal mobile devices (PMDs) in higher education is not new, access cannot be assumed in resource-constrained universities. We cannot expect that all students have access to ICTs, nor can we adopt the view that everyone has the digital skills they need for university (Brown, 2012). In the South African context, technologically immersed and savvy youth are in the minority and represent an elite, rather than a majority (Brown & Czerniewicz, 2010). Many students entering university are outsiders to the digital world, and can thus be described as "digital strangers" (Brown & Czerniewicz, 2010, p. 363).

This cautions against uncritical adoption of the concept of young students as "digital natives" as espoused by Prensky (2001) who wrote from a first-world context. In our context, many students rely on mobile phones as their primary means of internet access off campus (Czerniewicz, Williams, & Brown, 2009).

South African universities are therefore consistently faced with an economic and moral dilemma: as students come from diverse backgrounds and geographical locations, materials, technological capacities and access to ICTs cannot be assumed (Broekman, Enslin, & Pendlebury, 2002; Brown, 2014). In addition, while this begins to address the issues of physical access there is also the issue of epistemological access to educational technologies, which remains unsolved through the mere provision of devices (Magunje & Brown, 2013). In fact, within our faculty, the Centre for Higher Education Development (CHED), we operate under a mission of social redress as many of the students we support come from disadvantaged social backgrounds (CHED Mission, 2016).

The University of Cape Town (UCT), South Africa's oldest university, is one of Africa's leading teaching and research institutions. In 2013, UCT started the Flexible Learning Pilot Project, which involved mandatory one-to-one laptop access for 476 students in four undergraduate courses. In order to facilitate ubiquitous laptop ownership across these courses, financial aid students were provided with a laptop by the university. Funding remains an issue as the pilot continues into its second year and seeks to explore how having a personal mobile device (in this case a laptop) can create an enabling environment for the use of emerging technologies for teaching and learning (Brown & Pallitt, 2014). Defining innovation is challenging especially when what's innovative in one context is not necessarily innovative in another. Veletsianos (2010) takes a view of focusing less on specific affordances of technologies, rather defining emerging technologies in education more broadly, as "tools, concepts, innovations, and advancements utilized in diverse educational settings to serve varied education-related purposes" (2010, p. 3). This view has been adopted by others in our South African context who have investigated the perceptions of emerging context (Gachago, Ivala, Backhouse, Bosman, & Ng'ambi, 2013).

Personal Mobile Device Project

In 2013, the Department of Higher Education and Training in South Africa commissioned the UCT to investigate how the financial investment in a PMD, whether on the part of the university or students themselves, adds value to the teaching and learning experience (Brown, 2014). In this chapter we focus on the influence of PMDs on teaching practice, particularly in the case of educators who have not had experience in using educational technology in their practice.

We draw on a pilot implementation conducted at UCT in the Bachelor of Arts (BA) extended degree programme (EDP) within the Department of Humanities (HUM). The EDP is an alternative way to complete the general BA. Admission into this programme is open to South African students from previously disadvantaged backgrounds whose National Benchmark Test score(s) show that they have the potential to succeed, but who have not obtained the school-leaving scores necessary for mainstream admission (Faculty of Humanities, 2016).

The first year of this programme attempts to provide various types of support (e.g. writing support, tutorials, etc.) with specially designed models within the humanities. The two main introductory courses/models in the first semester which all five lecturers taught on were Language in the Humanities and Texts in the Humanities. Language in Humanities aims to assist students with reading texts and producing texts. It further gives students the tools to attempt to critically analyze texts—to identify the what, who, and how of texts—and it will provide them with examples and frameworks to help them create texts themselves. Furthermore, it attempts to explain the principles of argumentation and the meaning of discourse in order to assist them with the course, Texts in the Humanities (Faculty of Humanities, 2016).

As outsiders to the courses, we understood that we needed to have an insider who could liaise with the other lecturers on the courses about our project or about other matters related to the course. We had worked with one particular lecturer on a previous project and we maintained this relationship. We refer to her as the innovator here, as she was champion of the project.

Working through an innovator in order to demonstrate the possibility of pedagogical innovation, we were able to introduce BYOD (Bring Your Own Device) into this single course context, without requiring other educators to change their practices initially. Through researching the intervention and through lectures researching their own practice, we are able to demonstrate the potential of PMDs in the classroom. We aim to provide evidence from the particular context that we were able to scale up an innovation from one lecture and one class to five educators and one course among a group of educators, and thus develop a model for wider institutional implementation in a resource-constrained context.

Many students who enroll in this programme come from a low socio-economic background. We therefore selected this programme for the project because these students not only required additional academic support but a third of the class did not have the financial means to acquire any PMDs to further assist with their studies. Given this context, the course lecturers previously relied on computer laboratories for students and had developed a largely paper-based multimodal pedagogy. The opportunity offered through a new model of BYOD enabled them to consider how technology could be more purposefully integrated into the classroom.

We developed a short needs assessment survey (Figure 10.1). A total of 202 first-year students completed the survey during HUM's orientation week in January 2016. The results indicated that 70 students (35 per cent) did not have any type of PMD. These students were then informed via email and SMS that they would be provided with Proline tablets during the third week of the first semester to assist them with their studies. Because many of these students never had access to any device prior to receiving these tablets, we provided a basic training session around how to use the device as well as how to download university applications (VULA—our learning management system, email, etc.).

Framing Innovation and Adoption

There are many ways of trying to understand how change happens in the context of educational technology. Researchers have looked at how communities of practices influence adoption of elearning (Aceto, Dondi, & Marzotto, 2010; Chikh & Berkani, 2010), and also how innovation diffuses or moves through the system (Graham, Woodfield, & Harrison, 2013). In conceptualizing how innovation

Personal Mobile Device Ownership Survey

(Fill in the blanks. Tick the correct option)

This survey will assist us in understanding student's ownership of Personal Mobile Devices

Name: ____________________ **Student number:** ________________________

Cell phone number: ____________________

Where do you live? [ON Campus] OR [OFF Campus]

Other: ____________________

1. Do you have a **mobile phone (cell phone)**? [Yes] [No]

   Can you connect to the **internet** with your cell phone? [Yes] [No]

2. Do you have a **laptop**? [Yes] [No]

   Can you connect to the **internet** with your laptop?

3. Do you have a **tablet**? [Yes] [No]

   Can you connect to the **internet** with your tablet? [Yes] [No]

4. Would you be able to bring any of these device to campus every day? [Yes] [No]

   If no, why not? ____________________

FIGURE 10.1 Needs Assessment Survey

occurred in this case study, we have found Rogers's Diffusion of Innovation (DOI) theory useful (Rogers, 2003). Although it may seem a bit clichéd to talk about innovators and early and late adopters, DOI theory offers many useful ways of understanding what we observed happening in this context. What we draw on in this chapter are five qualities of innovation, the conditions or factors that need to be in place for an innovation to spread through a group. The first, Relative Advantage, is the degree to which the innovation is perceived to offer a benefit. This could be in terms of economic or social benefits. The second, Compatibility, refers to the synergy between the innovation and existing values and practices. The third, Simplicity, is about ease of use. The fourth, Trialability, indicates the extent to which the innovation can be experimented with in a limited capacity, i.e. the innovation offers less risk. Finally, the fifth is about the Presence of Observable Results (Rogers, 2003). Others have drawn on DOI to look at educators' adoption of elearning tools in a schools context (Jwaifell & Gasaymeh, 2013) as well as the relationship between DOI's attributes of innovation and intended and actual use (Ju Joo, Lim, & Lim, 2014). In this chapter we use DOI as a language of description to frame our interpretation of the behaviours and practices we observed. We also use the innovation-decision framework (Rogers, 2003) to frame changing practices and look at whether other educators in the course team chose to adopt the innovation or reinvent it.

Methodology

Over the course of April and May 2016, the project researcher (one of the authors of this chapter) interviewed the five lecturers on these courses to gain an understanding of their personal mobile use, their pedagogical practices specifically around technology use, and their perceptions on students using personal mobile devices in their classroom. The aim of the interviews was to reflect on the lecturers' prior pedagogical positions and practices at the start of the course and how they may (or may not) have changed in the duration of the semester.

During the interview process it became clear that one of the course team (whom we have since termed the innovator and is another co-author of this chapter) was leading the pedagogic practices around the use of the PMDs, and these practices were either taken up by the other lecturers or got them to begin thinking about more innovative ways to teach on other courses.

In order to understand the influence of this innovator in mainstreaming new pedagogic practices, we, the researcher and project leader (the other co-author), undertook a thematic analysis of the transcripts (Marks & Yardley, 2011). We coded deductively using the DOI factors as themes in Nvivo and focused on the qualitative meanings that emerged from the five individual lecturers' experiences. The data presented is a story of five cases rather than an empirical account of changing practice.

Findings

Personal Mobile Device Use by Academics

Before looking at how how the pedagogic practice began to change due to an innovator, we need to understand what the lecturers' ideas and perceptions are around PMD use in and outside the classroom.

Questions posed to lecturers in this regard were:

- What personal mobile devices do you have? Cell phone, tablet, laptop?
- Are you comfortable with students using devices in your classrooms or do you prefer they not use them? Why?
- Do you make use of any form of technology in your classroom?
- What facilities do you have in your classroom to make using technology possible?
- What have some of the advantages and challenges been with using or not using technology in the classroom?

The responses to the above questions provided a sense of the basis from which they could develop their pedagogical practices. In terms of our innovator, it was interesting to see how she personally used mobile devices as this could have possibly influenced her incorporating a blended approach in the classroom. She describes her dependency on her device:

> I think we've all become quite dependent on devices … the smartphone that plays so many functions today. From checking emails quickly to WhatsApping, staying in touch with friends or having groups
>
> *Interview 1, 13th April, 2016*

Her use extends well beyond the personal and she has integrated her personal and professional practices in terms of her use of her cell phone:

> obviously it's also used for entertainment purposes, but I use it a lot for research purposes. And not just to access written, you know, journal articles … . I watch a lot of Ted Talks and … I just want to have a quick sense of what are the key debates happening in the world as we speak … . I think that then filters into my teaching as well because I can give students very current examples.
>
> *Interview 1, 13th April, 2016*

Our innovator is aware that being connected (via your smartphone) in her personal capacity is important because not only does it keep her connected to family and friends, it also has the potential to filter into her professional life as a researcher and educator.

In contrast, the rest of the teaching team reflected a wider range of personal experience and confidence with technology. Three of the lecturers make quite traditional uses of their cellular phones, dominated by calls, SMS and instant messaging. Interviewee 2 describes herself as "quite a basic cell phone user". She tends to keep her cell phone for more traditional activities: "I don't have Facebook on my cell phone, … I use it for WhatsApp, SMS and really occasional[ly] for some job stuff … I find it easier on a computer" (Interview 2, 2nd April, 2016).

Interviewee 3's use is also traditional in the sense that he describes it as "normal sort of telephone calls, SMSs, … WhatsApp. I do some minor searching on the phone, sort of Googling for information … . And other than that it's to find out where my daughters are and what they're … ."

He has made a strategic decision not to let his professional life intersect with his personal: "I don't … I've purposely not linked my emails up to my phone" (Interview 3, 13th April, 2016).

While Interviewee 4 describes his use of the cell phone as predominantly for communication, he uses it across work and personal contexts:

> with postgrad students, and social media such as WhatsApp … [it's] more about trying to organize meetings or times that are suitable. And it's now and again there'll be a discussion coming up on a concept or something that's theoretical. And then also for our language and humanities course we have a WhatsApp group as well, and there sometimes we discuss

> marking issues coming up while we're marking an essay, or marking the exam.
>
> *Interview 4, 14th April, 2016*

For three lecturers mobile technology, while part of their daily lives, tends to be used for traditional communication purposes. Interviewee 5 is more enthusiastic about his use of technology in all aspects of his life:

> So I'm addicted to my phone, I'm on my phone all the time. Obviously my laptop I use a lot. I have a tablet at home that I also use. Social media, looking for information, research; so kind of entertainment and academic functions.
>
> *Interview 5, 14th April, 2016*

However, despite these varied views, the teaching team were all amenable to participation in the project. Rogers (2003) describes this as an Innovation–Decision process. In this case the process was somewhat expedited. The innovator and researchers on the project team met with the course team to talk about what the project was about. The innovator, being a colleague, was able to inform the teaching team about the project from an insider perspective. The Knowledge Stage and Persuasion Stage were therefore intermingled as the implications of potential use were discussed and outlined. Given the desire for simplicity and ease of implementation, the decision to implement or accept the PMD project into the course was based more on principle than practice. The intention was not to disrupt the teaching team's existing pedagogical practices.

Blended Learning and Mainstreaming Pedagogical Practices

In this case study, relative advantage was evidenced in terms of the opportunity lecturers perceived it offered for more flexible and expanded opportunities for learning for students. Our innovator had previously experimented with educational technology interventions in her class and was keen for support from a project which addressed issues of equity of access among her students in a systematic way. She is the person who wanted students from the course she taught in the first year EDP involved in our PMD project: "This model was quite appealing, this blended model, and therefore in 2014 already the director of the Education Development Unit asked me to propose the model to departments in the humanities as well" (Interview 1, 13th April, 2016). She already had a good sense of the pedagogical innovations she could introduce as a result of the conditions of access enabled through the project.

Second, the innovation was compatible with the courses' existing values and practices, as educators already adopted blended learning which consisted of a multi-modal approach where students had three contact or traditional lecture sessions, usually with a particular theme, from Monday to Wednesday, followed by a session where students went to a computer lab on campus on a Thursday in order to complete an online task. The online task incorporated key concepts or theories discussed in the contact sessions earlier in the week, which was designed to enable students to critically engage with and apply the concepts or theories learnt in the contact sessions:

> The design principle is that students will be presented with additional multi-modal resources and content in order to stretch their understanding of what was covered in lectures that week. [And] make them apply what they've learnt.
>
> *Interview 1, 13th April, 2016*

The whole course team supported the multi-modal approach to learning. They had in fact been working along these lines, drawing on different sources and texts, before the PMD project but were limited to what could be shown during classroom time and presented on paper handouts or via the learning management system. Expanding the multi-modal approach online was therefore a natural extension, especially for students on the introductory courses. Although lab sessions enabled them to do this without disadvantage, the PMD project offered greater flexibility for students to engage with learning in their own way and at their own time, and to revisit what was being covered.

The intervention addressed the need for simplicity and ease of use as the innovator conceptualized and structured the blended learning activities and the rest of the course team did not have to do this. In the weekly face-to-face meetings, some lecturers would suggest ideas for weekly tasks based on what students were encountering in the classroom. As such, even if the course team was not directly involved in the online learning design, they could openly contribute to it, which in turn increased their belief and investment in the project. The intention was to create an enabling teaching and learning environment for students and the five lecturers in the two EDP courses without the lecturers having to change their teaching practices as a result of the project—we know that innovative pedagogy is often driven by individuals and in academic contexts it is not easy to persuade or change other lecturers' pedagogical approaches in traditional face-to-face settings.

The innovator developed the task material which is presented to students by means of VULA (our learning management system). This requires that each student be logged into their VULA account, hence the necessity of being in a computer lab for the lab session. Previously this was essential as the lecturers could not assume all students had access to technology. However, with the roll-out of our PMD project and the combination of BYOD and the provision of tablets for the remainder of the class this is no longer a necessity,

The innovator then provided an information session or workshop to the other four lecturers who undertook the lab session with their respective classes. She took them through the actual task on the learning management system to illustrate how the various concepts have been illustrated by using media images (e.g. print advertisements or video advertisements). The innovator had constructed these tasks in such a way as to embed questions into the actual images or video clips, promoting students' interaction with the resources, and offering lecturers the possibility of tracking students' progress through a growing repository of online responses and site statistics.

In terms of Trialability, the innovation did not present a risk to the educators. In this context while PMDs were provided for all students in the class to use for their learning across all their courses, there was no expectation that other courses needed to acknowledge or change their pedagogical practice.

The course team met regularly to discuss the teaching and classes. This peer discussion is an important component of adoption as it allowed reflection on what was working and what was not (i.e. observable results). Timely exchange of information and just-in-time problem solving meant there was less uncertainty among the rest of the teaching team. In addition, each individual lecturer was getting feedback from students in their classes, which gave them an immediate sense of how the PMDs and blended sessions were valued. The feedback looped back onto practice and was factored into forthcoming blended sessions through a process of reflective and responsive design.

Changing Practice

In this chapter we have focused on how through creating a model of equitable access to PMD we have changed the attitude to and adoption of educational technology practice among a group of lecturers. What this project highlights is how an innovator can lead by example and provide the scaffolding for new pedagogical practices that involved PMD learning, and how late adopters can participate in change without having to rethink or change their teaching approaches immediately. After participating in the project for a semester, the teaching team are in a position to decide whether to reject the innovation or confirm it. In our case none of our educators has taken a decision to discontinue the innovation in their teaching practice. However, what we see in the process of confirmation involves what Rogers calls Re-Invention, i.e. the process of adopting and adapting the innovation in a way to better meet the needs of the people involved and the pedagogical ends of the course.

This has come through in the various interviews where lecturers are thinking about how they can change and adapt their teaching to be more open to PMDs. For example, one lecturer is participating in the design of an online course, two are developing a MOOC (Massive Open Online Course), and another is undertaking research and writing about the practice as a way of thinking through it.

Despite Interviewee 2 initially indicating that she did not like using PMDs for either her personal or professional life, it is quite revealing how she has adapted her pedagogical practices through working on this project:

> I've got a digital component in [xxx course]. Not as well developed as the Humanities one ... but I am finding my way around that. ... Designing, an online course for Postgrads.
>
> *Interview 2, 2nd April, 2016*

Another is rethinking how they can change the classroom practices for their course next year:

> Because I'm just thinking now if I could be at the front of a class and be able to send, even if each student doesn't have a tablet but there are some students with tablets, an image. And then they can work in groups using that specific tablet looking at the image there and looking up information about the image. That to me is amazing because then it's the digital literacy, they're using technology in the classroom, but they're also working in their groups.
>
> *Interview 5, 13th April, 2016*

Interviewee 3 perhaps best sums up the project as he reflects on how there are possibilities that can still be realized and is more conscious of how PMDs are enabling learning and reflection for students outside of the classroom:

> it's still in the process of morphing into much more than it started out. What we do now, realizing all the possibilities that it holds in terms of our course, in terms of the application of the theory. Over and above it's also created a platform for students to do a different type of writing that they do in class, so like the reflective stuff that they do ... it has allowed us to actually meet students where they are, yes?
>
> *Interview 3, 13th April, 2016*

Conclusion

The aim of the overall project is to explore the effectiveness of BYOD in flexible learning in our resource-constrained context. In this chapter we have looked at what this means for a small group of teachers as they grapple with the challenges and opportunities this affords in terms of their own pedagogical practices. In a context such as ours, where access cannot be assumed and lecturers are used to working within the constraints of fixed time dependent ICT infrastructure, the flexibility for PMDs to support pedagogical aims both inside and outside the classroom offers all kinds of new opportunities for teaching and learning that previously could only be imagined. The educators felt that by enabling students' access to technological devices they were able to explore new forms of literacies and media to interact with the subject matter in a more engaging and tangible manner. We have shown that even when the jump to using educational technology has been big for lecturers both in terms of their own digital abilities and in terms of their preferred teaching practices, a low stakes intervention such as this can have huge benefits in terms of helping them reimagine possibilities, which they might have been averse to before the technology was formally introduced.

And the theory of DOI might potentially offer a model for change management, as the approach to encouraging adoption of innovation in use of PMDs in our context was organic and yet successful. The factors of keeping the technology simple, introducing it on a small scale, providing support from within the team, and continuously reflecting and discussing the innovation so that positive benefits were observable proved to be key elements in the acceptance (and even reimagining of the potential) by educators.

References

Aceto, S., Dondi, C., & Marzotto, P. (2010). Pedagogical innovation on new learning communities: An in depth study of twelve online learning communities. In K. Ala-Mutka & A. Ferrari (Eds.), *JRC scientific and technical reports*. Spain: Joint Research Centre, Available from http://www.oerafrica.org/system/files/8893/pedagogical-innovation-new-learning-communities_0.pdf?file=1&type=node&id=8893

Amory, A. (2014). Design and implementation of strategies and artifacts to support ubiquitous computing in and outside the classroom: A reflective case study. In J. Viteli & M. Leikomaa (Eds.), *Proceedings of EdMedia: World conference on educational media and technology* (pp. 432–442). Association for the Advancement of Computing in Education (AACE).

Bozalek, V., Gachago, D., Alexander, L., Watters, K., Wood, D., Ivala, E., & Herrington, J. (2013). The use of emerging technologies for authentic learning: A South African study in higher education. *British Journal of Educational Technology*, *44*(4), 629–638.

Broekman, I., Enslin P., & Pendlebury, S. (2002). Distributive justice and information communication technologies in higher education South Africa. *South African Journal of Higher Education, 16(1)*, 29–35.

Brown, C. (2012). University students as digital migrants. *Language and Literacy*, *14*(2). 41–61.

Brown, C. (2014). *Investigation into enabling conditions to optimise the use of personal mobile devices (PMDs) in teaching and learning in Higher Education Institutions in South Africa.* Proposal submitted to the Department of Higher Education Collaborative Teaching Development Grant Program.

Brown, C., & Czerniewicz, L. (2010). Debunking the "digital native": beyond digital apartheid, towards digital democracy. *Journal of Computer Assisted Learning* *26*(5), 357–369.

Brown, C., & Pallitt, N. (2014). *Laptops and learning spaces: Online, offline and in between.* Network Learning Conference, Edinburgh April 2014.

CHED Mission, (2016). Retrieved from http://www.ched.uct.ac.za/ched/chedmission

Chikh, A., & Berkani, L. (2010). Communities of practice of e-learning, an innovative learning space for e-learning actors. *Procedia – Social and Behavioral Sciences, 2*(2), 5022–5027.

Czerniewicz, L., & Brown, C. (2009). A study of the relationship between institutional policy, organisational culture and e-learning use in four South African universities. *Computers and Education, 53*(1), 121–131.

Czerniewicz, L., Williams, K., & Brown, C. (2009). Students make a plan: understanding student agency in constraining conditions. *ALT-J Research in Learning Technology, 17*(2), 29–35.

Faculty of Humanities, (2016). *Undergraduate degrees and programs*. Retrieved from http://www.humanities.uct.ac.za/hum/undergraduate/degrees.

Gachago, D., Ivala, E., Backhouse, J, Bosman, J.P., Bozalek, V., & Ng'ambi, D. (2013). Towards a shared understanding of emerging technologies: Experiences in a collaborative research project in South Africa. *The African Journal of Information Systems, 5*(3). Available from http://digitalcommons.kennesaw.edu/ajis/vol5/iss3/4

Graham, C., Woodfield, W., & Harrison, J.B. (2013). A framework for institutional adoption and implementation of blended learning in higher education. *Internet and Education, 18*, 4–14.

Ju Joo, Y., Lim K., & Lim E. (2014). Investigating the structural relationship among perceived innovation attributes, intention to use and actual use of mobile learning in an online university in South Korea. *Australasian Journal of Educational Technology, 30*(4), 427–439.

Jwaifell, M., & Gasaymeh, A. (2013). Using the Diffusion of Innovation Theory to explain the degree of English teachers' adoption of interactive white boards in the modern systems school in Jordan: A case study. *Contemporary Educational Technology, 4*(2), 138–149.

Magunje, C., & Brown, C. (2013). From cellphone to computer: University students' use of technology in first year. In E. Ivala, (Ed.), Proceedings of the 8th International Conference on e-Learning. *Reading: Academic Conferences and Publishing International Limited*, 496–502.

Marks, D. & Yardley, L. (2011). *Content and thematic analysis. In research methods for clinical and health psychology*. London: Sage Publications.

Prensky, M. (2001). Digital natives, digital immigrants: Part 1. *On the Horizon, 9*(5), 1–6.

Rogers, E. (2003). *Diffusions of innovations* (5th ed.). New York: Free Press.

Veletsianos, G. (2010). *Emerging technologies in distance education*. Edmonton, AB: AU Press. Retrieved from http://www.aupress.ca/books/120177/ebook/99Z_Veletsianos_2010-Emerging_Technologies_in_Distance_Education.pdf

11

MOBILE LEARNING TRANSFORMATION IN A NATIONAL HIGHER EDUCATION SYSTEM

Cathy Cavanaugh and Jace Hargis

Introduction

Digital transformation is adopted as a strategy in organizations globally and in all sectors to prepare for "digital disruption" (Kane, Palmer, Phillips, Kiron, & Buckley, 2016). Education as an enterprise is in the bottom five in having a clear strategy and skilled leaders for digital transformation (Kane, Palmer, Phillips, Kiron, & Buckley, 2015), two essentials for successful transformation. Further, "success of every initiative is dependent on how well its implementation is aligned with a shared vison, mission and goals, that truly reflect the role of education in the new context of a digitally-rich world" (Dixon & Einhorn, 2016, p. 21).

This chapter examines a higher education system's digital transformation by highlighting its vision, strategy, leadership and research on initial outcomes. Few comprehensive system-scale, evidence-based stories are available for higher education digital transformation. The transformation case study is a synthesis of multiple published studies focused on the year encompassing the planning and initial implementation of new technology in all institutions within a country's higher education system.

In April 2012, ministry-level higher education leaders in the United Arab Emirates (UAE) charged the three federal higher education institutions, comprising 20 campuses, to create meaningful mobile learning (mlearning) programs in and outside of the classrooms, in part by providing tablets to all students and instructors along with digital curriculum. At the time, this program represented the largest education iPad initiative in the world, according to Apple education executives (McCarthy, 2013). The initiative emphasized that pedagogy, rather than technology, would guide the implementation and by September 2012 the campuses would fully transform their learning environments to a mobile learning ecosystem, and teaching practice would evolve to increase engagement and student-centered learning experiences for all incoming students. The higher education mobile learning initiative was part of the National Strategic Priorities that included First Rate Education with "smart systems and devices," and a Competitive Knowledge Economy that promotes innovation and entrepreneurship (UAE Cabinet, 2011, n.p.). Since that time, the UAE's National Innovation Strategy was enacted to develop capacity for a broader entrepreneurial economy. Included in the National Innovation Strategy were "innovation labs in schools and universities as part of a drive to equip students with targeted skills such as critical thinking, problem-solving, creativity, perseverance and adaptability" (UAE Cabinet, 2014, n.p.).

Literature/Theoretical Background

The UAE's higher education transformation to mobile learning environments in support of meaningful student-centered learning is a story of organizational transformation enabling classroom transformation. Models of organizational transformation in business (Kane, Palmer, Phillips, Kiron & Buckley, 2016) and education (Cavanaugh, McCarthy, & East, 2014; CoSN, 2015; Kools & Stoll, 2016) agree on several key characteristics of digitally mature learning organizations: leadership, vision, collaboration, capacity building, partnerships and strategy for a holistic change. These digital transformation frameworks are underpinned by theories of organizational congruence (Nadler & Tushman, 1980; Cameron & Quinn, 2011). Essentially, in an organization "congruence predicts success" (Cameron & Quinn, p. 53). Education systems and schools in which core elements are aligned are more likely to avoid distractions and remain focused on quality. Figure 11.1 shows the core elements of organizational congruence from models of digital and educational transformation and how they relate to the UAE mobile learning initiative. The remainder of the chapter will summarize the research studies that examined the initiative.

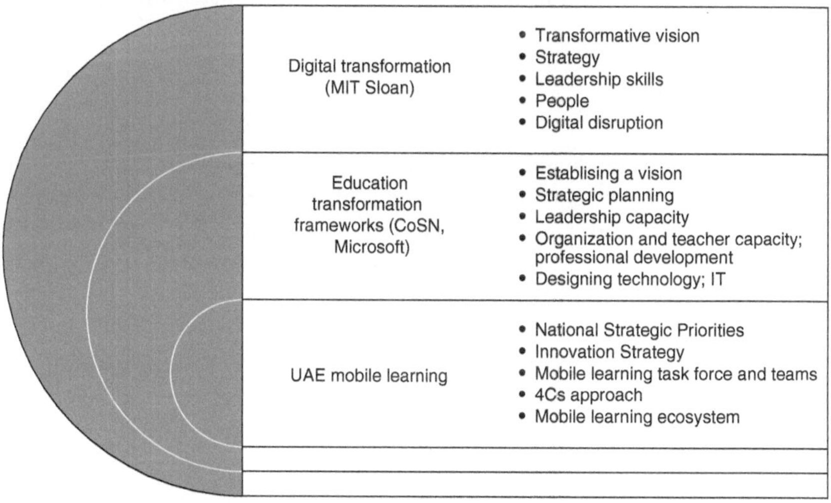

FIGURE 11.1 Transformation Models for Organizational Congruence.

With attention at the highest levels of leadership on the keys to system transformation, the context is enriched for transformation of the teaching and learning experience, which has shifted in higher education toward deeper, more relevant approaches (Johnson, Adams Becker, Cummins, Estrada, Freeman, & Hall, 2016) that depend on active, authentic and meaningful pedagogies Newmann, 1996; Jonassen, 2008). Tablet devices with digital content constitute an extensible, cognitive toolbox for authentic, active learning centered on interaction among students, faculty members and community experts (Cavanaugh & Hargis, 2013b).

The new mobile learning ecosystem facilitated authentic and meaningful engaged learning all the time, everywhere (Cavanaugh & Hargis, 2013a). It was uniquely suited to support the essential learning actions of monologue, dialogue and trialogue: reflection, conversation, interaction with tools and media (Cavanaugh & Hargis, 2013b). The blended learning ecosystem that was built purposely around students with mobile technology increased the roles and importance of the instructor who iteratively designed the learning experience from a very wide and continually shifting array of possibilities (Clark & Barbour, 2015).

Method

During the initial year of the UAE's higher education mobile transformation, adoption of the technology, changes in pedagogy and patterns among learners were documented. Participants in the research in the initial year comprised the three groups who were most engaged in the digital transformation:

1. Faculty mobile learning leaders at each campus
2. Faculty instructors in the programs receiving new students
3. Students who learned in the pilot blended courses offered that year in one college

Champions

Faculty leaders were identified as innovative educators on their campuses who would learn the new technology tools and facilitate learning of their colleagues. Thirty of these leaders across the 20 campuses participated in a 4Cs approach to collaborative faculty development (Cavanaugh, Maor, & McCarthy, 2014) in which:

1. They were designated as Champions in their campuses and participated in the intensive faculty development focused on mobile pedagogy.
2. They Created new learning experiences for their courses.
3. They Communicated with each other in a supportive virtual community and with their campus peers to share new student-centered active approaches.
4. They led the nationwide Celebration events that brought all faculty together for networking and learning twice each year.

These Champions became primary drivers in the success of the mobile pedagogy project, and they engaged in personal or structured reflection on their mobile teaching (Hargis, Cavanaugh, Kamali, & Soto, 2013a, 2013b). Nineteen of the 30 completed data collection via their online collaboration environment. Four Champions were purposively sampled for a comprehensive survey of Technological Pedagogical Content Knowledge (TPACK) (Koehler & Mishra, 2009).

Instructors

Faculty instructors in each campus learned from their Champions in a range of structured and unstructured activities, peer observations, collaborative lesson development and the Celebration events at which hundreds shared their innovative approaches (Cavanaugh, Hargis, Kamali, & Soto, 2013; Cavanaugh, Hargis, Munns, & Kamali, 2013). The lessons that instructors shared in these events were coded by Substitution, Augmentation, Modification or Redefinition (SAMR) level at each event (Puentedura, 2012). In addition, faculty could elect to join teaching and learning teams to closely examine their new practices. Across the country 224 of the 325 eligible instructors completed a survey of attitudes toward technology-supported learning environments, the Faculty Adoption of Technology Supported Learning Environments (FATSLE), which was adapted for this study from the Knezek et al. (1999) Faculty Attitudes Toward Information Technology (FAIT) survey that had been adapted and rewritten by Tinnerman (2007) and further adapted for the blended environment by Jordan (2010). In addition, the survey included two items adapted from Christensen (1997) to assess faculty stages of adoption of technology and their phase technology integration. A purposive sample of three instructors was observed by clinically trained education experts using an instrument adapted for this use from the "Teacher Observation Report of Student Centered, Inquiry-Based Classroom Instruction" originally developed for a U.S. National Science Foundation-funded professional development program (Greater Philadelphia Secondary Mathematics Project, 2005). Faculty instructors at one college piloted an experiential learning project as an instantiation of active engaged student-centered professional learning by sharing their learn-by-doing approaches in a community event (Cavanaugh & Hargis, 2014). These approaches were reviewed using Newmann's rubrics for authentic instruction as an indicator of the levels of higher order thinking, connection to the world beyond the classroom, and depth of knowledge present in the newly adopted activities.

Students

During the mobile learning transformation year, a pilot blended course involving seven instructors and 477 students was offered on one campus, a women's college (Cavanaugh, Mayberry, & Hargis, 2016). The course was a liberal studies course, required for all students and not a course in the professional degree major field. Between one-third and one-half of the course activities were completed in an online learning management system (LMS), reducing in-class meeting time. Students' final course grades were used as the measurement for student success. Data logs generated by the LMS during the course included the number and length of student logins to the online course components.

By examining a range of data from these three key participant groups, we could characterize the nature of organizational digital transformation in the UAE. Data were collected throughout the planning and implementation year, including classroom observations using a rubric for active mobile learning, interviews with faculty, focus groups and reflections from Champions and sample lessons using mobile technology.

The following table summarizes the data collection strategy.

Participants	*Timing/Frequency*	*Data Source*	*Purpose*
Campus Champions across the UAE N = 19	Initial month of implementation	Focus groups and reflections	Perceptions of the initial stages of the initiative
Campus Champions on different campuses N = 4	Initial month of implementation	Survey of SAMR and TPACK	Assess levels of technology adoption and integration
Faculty instructors at all campuses N = 224	Start of Semester 1	Survey of attitudes toward blended learning environments	Assess baseline attitudes across instructors
Faculty instructors at all campuses N = 81	Prior to start of classes, end of Semester 1, end of Semester 2	Sample lessons	Depict change in practice over the initial year
Faculty instructors at two campuses N = 3	10 weeks into Semester 1	Observations of class teaching	Snapshot of practice
Blended course students on two campuses N = 477	End of course	LMS log data, course final grade	Insight into characteristics of successful students

Findings

A leader's job is not to design a change program but to build a change platform "that allows anyone to initiate change, recruit confederates, suggest solutions, and launch experiments" (Hamel & Zanini, 2014, n.p.).

The goal of national education leaders and higher education campus leaders during the UAE's mobile learning transformation was to create the conditions for increased active, engaged, student-centered learning as part of the national strategy to enact the country's education vision, which depended first on building leader and educator capacity. The varied approach to data collection among three key stakeholder groups in the initial year of the initiative constitutes a comprehensive window into the perceptions, experiences and shifts, and indicated whether the transformation was approaching its purpose. Results for each group are summarized here.

Champions

The campus Champions participated in focus groups and reflections during the early stages of the initiative, and a small sample completed a survey focused on TPACK to indicate levels of technology integration and adoption. Focus groups showed that the new technology was used quickly by students, and faculty were creating new teaching approaches, although they were on a simple level initially. Among the challenges reported were somewhat cumbersome workflow connecting teaching materials and student work and feedback, and dissatisfaction with quality of digital textbooks. The survey showed that each participant reached an accomplished level in all areas of TPACK.

Instructors

Instructor surveys of attitudes toward blended learning environments showed that as the program was beginning most respondents rated their technology adoption and integration between Understanding and Fusion levels, which is a strong starting point at the launch of a transformation:

- Understanding and application of process—I am beginning to understand the process of using technology and can think of specific tasks in which it might be useful.
- Fusion—I can use more than one application in the creation of a single product. I use technology including iPads in preparation, instruction and evaluation. My students use a variety of applications regularly in the construction of curriculum-based products.

In addition, on specific items, 80 per cent of instructors reported that their most frequent use of technology was to promote student-centered learning; as a communication tool, 95 per cent felt their administration actively encouraged the use of the technology in classrooms; and 73 per cent were satisfied with campus support. Areas for growth included adequate training, preparation and confidence.

Instructor observations using the student-centered mobile learning protocol indicated strong skill (above the mean) in facilitating lessons that actively involved students in the lesson introduction through participation in an opening activity, eliciting prior knowledge, accommodating a range of learning styles, using the technology to enhance the activity, and using small group discussions that were integral to the objective. They also showed strength in using time for reflection and discussion, balancing group and individual accountability, providing feedback, and redirecting questions to students. The lesson samples were rated lower on student enthusiasm and excitement about the lesson. In the observed lesson samples, there was far more student discussion than teacher talk, and students were all actively engaged in a task on the mobile technology within a group. The most striking result was seeing that each lesson made use of multiple mobile apps and media resources in the lesson, as well as multiple forms of communication in the application of the lesson content and multiple forms for evidence of learning.

Lesson samples collected from instructors at three points during the initial mobile learning transformation year as part of the Celebration sharing events were rated by trained raters for SAMR levels. A chi-squared analysis of descriptions of teaching practices was used. The data indicate faculty-driven practice-embedded development and intentional integration of the two models were helpful in changes in teaching practice in a short period of time. Change was observed in pedagogy, technology and content indicators among the faculty who shared their mobile teaching practices. Faculty showed adoption and implementation of new tools, resources and practices can begin quickly and at large scale.

Instructor approaches to experiential learning were collected in the second half of the mobile learning initial year in a college-wide sharing event and rated by trained raters according to levels of authentic learning. Results of analysis indicate the sharing event was effective in expanding faculty knowledge and adoption of experiential mobile approaches in their courses as evidenced by growth in specific approaches used by faculty in the semester following the sharing event. The ideas shared by faculty at the outset of the national mobile learning initiative were rated high in indicators of higher-order thinking, moderate in indicators of depth of thinking and low in indicators of connection to the world beyond the classroom. As faculty continued working with the new technology, they learned more possibilities for extended complex projects, and for connections beyond the campus.

Students

At the conclusion of a blended learning course in one college, a regression analysis was used to examine relationships between student use of the LMS and their final course grade. LMS data included numbers of logins, accumulated time logged in, and average session length. Logins appear to be a more significant predictor than average session length and course grade is predicted to be highest at intermediate values of both predictors. Thus, too many or too few logins and too little or too much time in the LMS were not associated with better grades.

Conclusion

During the year of mobile learning transformation, the UAE's federal higher education system showed the characteristics of an organizationally congruent learning organization: leadership, vision, collaboration, capacity building, partnerships and congruence or alignment of digital strategy within a holistic change approach. The nation's highest leaders advocated publicly and through policy for an innovative society, and national education leaders initiated mobile learning to put in place the conditions for a more innovative and engaged learning experience. Across the national higher education system, each campus was represented in leadership teams for pedagogy, content technology and research that were overseen by a central committee. Campus leaders and faculty collaborated with key partners on faculty development activities and research throughout the implementation year to build local capacity and document change. Thus the UAE's higher education system worked quickly to increase organizational congruence to accelerate its digital transformation. While many contextual factors may not be shared between the UAE and other higher education systems, the core elements of system alignment and transformation appear to apply generally.

Student-centered teaching practice was the core goal of the UAE's program, and evidence of success as well as some limitations were found, due to transformation leadership and commitment to ongoing practice-embedded faculty development efforts centered on active, engaged learning by doing. Faculty perceptions and uses of the tools and ecosystem indicated that faculty have quickly transcended basic substitution of analog for digital materials and made substantial progress at scale early in the initiative toward transformation of the learning experience. Students in the first blended learning course helped faculty and leaders to understand that blended courses can expand learning time—as shown in the range of logins and session lengths recorded in the course—and that effective blended learning courses can be designed to alternate physical and virtual experiences.

The research reported here demonstrated that system-wide education transformation is possible within a year with explicit cross-organizational attention to alignment across core elements of transformation. It also shows that illuminating and understanding the nature of transformation through a varied data collection strategy is an intricate and deliberate process. In the years since this research was conducted, advances in analytics have dramatically streamlined the process of gaining insights into education change and learning experiences. It is most actionable in education when we have just-in-time data and forecast trends. The scope of our research and ability to take action based on insights would be accelerated with advanced analytics technology (Cavanaugh, Sessums, & Drexler, 2015). Scaling research in mobile learning is crucial to advancing understanding as learning environments and pedagogies shift rapidly. Leaders and educators are encouraged to leverage the power of such tools in their practices.

References

Cameron, K., & Quinn, R. (2011). *Diagnosing and changing organizational culture based on the competing values framework* (3rd ed.). San Francisco: Jossey-Bass.

Cavanaugh, C., & Hargis, J. (2013a). An engaged and engaging mobile learning ecosystem: early findings from a large-scale college iPad program. *International Journal of Instructional Technology and Distance Learning, 10*(10), 33–42.

Cavanaugh, C., & Hargis, J. (2013b). iPads as cognitive toolboxes in higher education. In S. Dowling, ed., *Redefining Learning*, book 2 of the HCT Educational Technology Series (pp. 1–13). Abu Dhabi: HCT Press.

Cavanaugh, C., & Hargis, J. (2014). College "learn by doing" ideas picnic: Faculty-led faculty development for authentic instruction. *Journal of Transformational Learning, 2*, 47–64.

Cavanaugh, C., Hargis, J., Kamali, T., & Soto, M. (2013). Substitution to augmentation: Faculty adoption of iPad mobile learning in higher education. *Interactive Technology and Smart Education, 10*(4), 270–284.

Cavanaugh, C., Hargis, J., & Mayberry, J. (2016). Participation in the virtual environment of blended college courses: An activity study of student performance. *The International Review of Research in Open and Distributed Learning* [S.l.], *17*(3), ISSN 1492-3831. Retrieved from http://www.irrodl.org/index.php/irrodl/article/view/1811/3696

Cavanaugh, C., Hargis, J., Munns, S., & Kamali, T. (2013). iCelebrate teaching and learning: Sharing the iPad experience. *Journal of Teaching and Learning with Technology, 1*(2), *1–12*. Retrieved from http://jotlt.indiana.edu/article/view/2163

Cavanaugh, C., McCarthy, A. & East, M. (2014). An innovation framework for holistic school transformation: Ten critical conversations for the 21st century. In J. Viteli & M. Leikomaa (Eds.), *Proceedings of EdMedia: World Conference on Educational Media and Technology 2014* (pp. 1560–1569). Association for the Advancement of Computing in Education (AACE).

Cavanaugh, C., Maor, D., & McCarthy, A. (2014). Mobile learning. In R. Ferdig & K. Kennedy (Eds.), *Handbook of research on K–12 online and blended learning* (pp. 391–441). Pittsburgh, PA: Entertainment Technology Center Press, Carnegie Mellon University.

Cavanaugh, C., Mayberry, J., & Hargis, J. (2016). Participation in the virtual environment of blended college courses: An activity study of student performance. *International Review of Research in Open and Distance Learning, 17*(3).

Cavanaugh, C., Sessums, C., & Drexler, W. (2015). A call to action for research in digital learning: Learning without limits of time, place, path, pace … or evidence. *Journal of Online Learning Research, 1*(1) 9–15.

Christensen, R. (1997). Effect of technology integration education on the attitudes of teachers and their students. (Unpublished doctoral dissertation). University of North Texas.

Clark, T., & Barbour, M. (2015). *Online, blended and distance education in schools*. Sterling, VA: Stylus and Microsoft.

Consortium for School Networking. (2015) *Framework of essential skills of the K–12 CTO*. Washington DC: Author.

Dixon, B., & Einhorn, S. (2016). *Anytime anywhere learners*. Bellevue, WA: Anytime Anywhere Learning Foundation.

Hamel, G., & Zanini, M. (2014). Build a change platform, not a change program. Retrieved from http://www.mckinsey.com/business-functions/organization/our-insights/build-a-change-platform-not-a-change-program

Hargis, J., Cavanaugh, C., Kamali, T., & Soto, M. (2013a). A federal higher education iPad mobile learning initiative: Triangulation of data to determine early effectiveness. *Innovative Higher Education, 39*(1), 1–13.

Hargis, J., Cavanaugh, C., Kamali, T., & Soto, M. (2013b). Measuring the difficult to measure: iPad mobile learning. *International Journal of Mobile and Blended Learning, 5*(2), 60–77.

Jordan, D. (2010). *Faculty attitudes toward online and blended learning (FABOL).* (Unpublished survey). Stockton, CA: University of the Pacific.

Kane, G., Palmer, D., Phillips, A., Kiron, D., & Buckley, N. (2015). Strategy, not technology, drives digital transformation: Becoming a digitally mature enterprise. *MIT Sloan Management Review.* Retrieved from http://sloanreview.mit.edu/projects/strategy-drives-digital-transformation/

Kane, G., Palmer, D., Phillips, A., Kiron, D., & Buckley, N. (2016). Aligning the organization for its digital future. *MIT Sloan Management Review.* Retreived from http://sloanreview.mit.edu/projects/aligning-for-digital-future/

Knezek, G., Christensen, R., Gilmore, E. L., Kim, H., Magoun, A. D., Owens, C., Morales, C., Moonen, B., & Voogt, J. (1999). Teacher and student attitudes toward information technology in four nations. *Society for Information Technology & Teacher Education International Conference, 1*, 916–918.

Koehler, M., & Mishra, P. (2009). What is technological pedagogical content knowledge (TPACK)? *Contemporary Issues in Technology and Teacher Education, 9*, 60–70.

Kools, M., & Stoll, L. (2016). WHAT MAKES A SCHOOL A LEARNING ORGANISATION? Education Working Paper No. 137. Paris: OECD.

Johnson, L., Adams Becker, S., Cummins, M., Estrada, V., Freeman, A., & Hall, C. (2016). *NMC Horizon Report: 2016 Higher Education Edition.* Austin, TX: The New Media Consortium.

Jonassen, D. H. (2008). *Meaningful learning with technology*. Pearson/Merrill Prentice Hall.

McCarthy, A. (2013). Personal communication.

Nadler, D., & Tushman, M. (1980). A congruence model for organizational assessment. In E. Lawler, D. Nadler, & C. Cammann (Eds.), *Organizational assessment: Perspectives on the measurement of organizational behavior and quality of working life.* Hoboken, NJ: Wiley.

Newmann, F. M. (1996). *Authentic achievement: Restructuring schools for intellectual quality.* San Francisco: Jossey-Bass.

Puentedura, R. (2012). Building on the SAMR model. Unpublished essay.

Tinnerman, L. (2007). University faculty expressions of computer self-efficacy and personal attitudes regarding the viability of distance learning. (Unpublished doctoral dissertation). Indiana, PA: Indiana University of Pennsylvania.

UAE Cabinet. (2011). USA strategy 2011–2013. Abu Dhabi: Author.

UAE Cabinet. (2014). National innovation strategy. Abu Dhabi: Author.

12

KAKAOTALK MEETS THE MINISTRY OF EDUCATION

Mobile Learning in South Korean Higher Education

Michael S. Gallagher

Introduction: South Korean Higher Education and Mobile Learning

South Korea presents a rich environment for investigating mobile learning (mlearning) in higher education. It is a nation with a passionate educational legacy evidenced by 67.7 per cent of 25–34 year olds obtaining some form of tertiary education coupled with a robust local mobile industry, evidenced by mobile manufacturers such as Samsung and a sophisticated domestic mobile ecosystem. It is a nation where economic and educational volatility is being threatened by a vastly diminishing birthrate, which at 1.21 children per woman is the lowest among the Organisation for Economic Co-operation and Development member nations (OECD, 2014); higher education is reconfiguring and technological alternatives are being pursued and implemented in response.

Higher education in South Korea is itself a sophisticated orchestration of sociocultural factors heavily influenced by its historical legacy. It is a nation which has existed in proximity to, been subjugated by, and emerged independently from two neighboring countries. From China, South Korea has adopted Confucianism and drawn from it a highly structured social order which informs much of the structure of higher education itself; from Japan, South Korea identified economic and development models that allowed it to emerge from devastation at the end of a brutal war. Since 1910, Korea has gone from Confucian kingdom to Japanese colony to free state to divided state to civil and proxy war, poverty, rapid industrialization, and to its current manifestation as economic power. Several universities emerged from the post-colonial state (1945) built on Western models of higher education. As such, South Korean higher education is a complex orchestration of foreign organizational models built on indigenous Korean sociocultural foundations. While caution must be applied in overemphasizing the dichotomy of Eastern (typified here by South Korea) and Western (typified here by the UK) learning and practice, particularly due to their significant overlap and undertheorized nature (Ryan & Louie, 2007), differences remain. Students in higher education live in this space of Eastern and Western overlap.

Pressure is being placed on South Korean higher education as universities are closing based on the diminishing pool of students, the consequence of a declining birthrate. Eleven universities were closed between 2000 and 2014 out of a total of 220, with a further 47 universities having merged with 41 other universities (Korean Council for University Education, 2016). A further 11 universities have had their access to government funding curtailed, making their closure possible if not likely (Chosun.com, 2016). According to the Ministry of Education's Policy Office report (2016), there is the expectation that by 2018 university entrance quotas will outstrip the number of eligible high-school graduates, a divide that will continue to accelerate (by 2024 this divide will be 30 per cent greater than it was in 2014). The Ministry of Education (MOE) has suggested that this will devastate the nation's higher education system as it currently stands.

If higher education in South Korea is consolidating because of declining birthrates and shifting policy priorities, the domestic mobile industry and mobile consumption are not. With a technological infrastructure on par with any nation in the world, where 40 per cent of its population enjoy a high speed, fixed internet connection, and where there are over 111 mobile subscriptions per 100 people, South Korea enjoys an almost complete saturation of connectivity (ITU ICT, 2015). It is an environment of increasing mobility as mobile traffic accounts for 29 per cent of all web traffic (We Are Social, 2015). 30 per cent of the South Korean population actively using social media and 27 per cent of the entire population does so through mobile devices (We Are Social, 2015). Along with this access comes an array of mobile and media practices that present unique sociocultural characteristics. It is also from this environment that we begin to see sophisticated instances of informal mobile learning, of longstanding and iterative informal mobile learning communities, and of South Korean communicative practices. Mobile technology sits at the nexus of these community memberships as South Korean university students use it to manage relationships and identities across the layers of activity that these memberships demand. Upon entering South Korean higher education, these students have developed and honed practices over a course of years that make this management possible. South Korean higher education, in response to the demands of consolidation, has developed a formal policy for the appropriate use of ICT, including mobile technology, in a larger ecosystem of smart learning, which will be discussed further in this chapter. The informal meets the formal, problematically, through mobile technology.

Method

The methods employed for this thesis include desk research, particularly an analysis of pertinent policy documents from the MOE in South Korea, demographic information from the Korean Statistical Information Service (KOSIS), as well as research on informal mlearning communities and cultures in South Korea, most notably in Hjorth (2009a, 2009b, 2008), Yoon (2003), and Ok (2011). This desk research was conducted specifically to ascertain how the evolving higher educational landscape, one consolidating because of low birthrates, is affecting the ICT and mobile learning efforts of South Korean universities, and how those efforts cohere or conflict with informal mobile learning communities at work in larger South Korean society.

Further methods draw on, but do not directly reference, qualitative research conducted by the author on mobile learning in South Korean higher education. Specifically, select findings emerge from qualitative studies conducted by this author in 2013–2015 with graduate students in South Korean universities to determine the informal mobile learning communities in which they participate and how that mobile activity shaped participation in their university discipline. The author was affiliated with a South Korean university as an assistant professor, although none of the findings presented in this chapter emerged from data taken from the author's own students.

Current State of Mobile Technology Use in South Korea: The Formal

Partly because of the consolidation required by a diminishing pool of university age students, partly because of policy directed towards educational competitiveness, and partly as a result of the sheer ubiquity and performance of an ICT infrastructure, higher education in South Korea is experiencing rapid transformation, a transformation that directly affects the state of mobile learning in universities.

The MOE has contributed explicitly to this process in terms of both policy and infrastructure; in 2015, MOE support for bandwidth equated to approximately 10.5 per cent of its entire budget (KERIS, 2015, p. 109). For education, this saturation of broadband access, ICT and mobile technology has produced an environment where educational policies and programs developed around this technology are part of a larger interrelated whole. This whole generally falls under the aegis of the preferred nomenclature of the MOE, smart learning, adopted in South Korea by the precursor to the MOE, the Ministry of Education, Science, and Technology (MEST), in 2011 to support digital textbook development and application, online education, open educational resources and content, online assessment, and to strengthen information and communication, teacher's use of ICT, and more (Kwon et al., 2013, p. 10).

Smart learning incorporates mobile technology as an ancillary technology in a larger learning system; it encapsulates the environment and culture of seemingly unlimited internet access and technological ownership that most South Koreans enjoy. The objectives of higher education in South Korea, encoded in many government-sponsored directives and university curricula, have been redrawn considering the perceived advantages of technology-enhanced learning. These objectives include pedagogical shifts towards collaborative learning and the employ of ICT and mobile technology to enact these shifts. Yet, the South Korean smart learning model still assumes, even centralizes, measures of institutional or instructional control. In Korean higher education, mobile technology serves a largely supportive role, a role that is contrasted to the central role of mobile technology in the informal space. Mobile learning as positioned in this larger ICT environment in higher education generally ascribes to the following distinctions.

Support for Smart Learning: Mobile as Node in a Larger Technological System

Much of this type of mobile activity is directed at providing a seamless learning experience through making instructional content available through mobile technology. In this instance, mobile technology plays a supportive role as a means of accessing content and performing administrative duties, rather than for socialized interaction or peer-based learning. Mobile learning as positioned here involves accessing video lectures, course materials, and lectures. Relevant examples of this approach include leading science institutes such as the Korea Advanced Institute of Science and Technology (KAIST), Ulsan National Institute of Science and Technology (UNIST), and Korea National Open University (KNOU), as well as universities dedicated to online education: Seoul Digital University, Hanyang Cyber University, and so forth (Choi, Woo, & Jung, 2013).

Support for Cyber Universities

Continuing from this smart learning distinction are the digital and cyber universities themselves, terms that South Korea uses for its universities which are exclusively or predominantly online; these are also referred to as distance education universities as per the Higher Education Act of South Korea (MEST, 2009). There are 19 cyber universities in South Korea (as of September 2015), all bound to the Higher Education Act and thus MOE oversight. These cyber universities have grown modestly since their inception in the early 2000s, despite general reservations about the quality or equivalency of the education offered compared with more traditional universities (detailed in Choi, Woo, & Jung, 2013). As these universities are major stakeholders and participants in the smart learning projects encouraged by the MOE, mobile technology plays a supporting role in their education, limited mainly to course content and administrative functions. Some cyber universities have attempted to shift the bulk of their learning management systems to mobile "campuses" where interactivity includes attending online courses, posting questions, checking messages, and monitoring academic calendars. As such, we see a potential expansion of mobile technology's role in higher education. However, difficulties arise when students choose not to adopt the mobile learning system, due to its complexity and a general social resistance (Han & Han, 2014); mobile acceptance in higher education, and the methodologies designed to measure that acceptance, have been a popular research thread (discussed in the broader mobile sense in Ho Cheong & Park, 2005 and specifically to mobile learning in higher education in Park, Nam, & Cha, 2012).

Open Courseware and Open Learning Support

Higher education institutions and the South Korean government have invested in the development of Korea Open CourseWare (KOCW), modeled on MIT's open courseware, which is designed to utilize lectures and lecture data that Korean universities have made openly available. "It is an open knowledge resource hub that can collect, distribute and utilize high quality lecture contents diffused through universities in collaboration, and attempts to correspond to the flow of Open Educational Resources (OER) that is unfolding worldwide" (KERIS, 2015, p. 111). Mobile technology in this instance is used to support these efforts by providing media support services for the content and courseware, and by composing location-specific automated feedback systems to activate lecture openings. Furthering this trend towards open learning are K-MOOCs, which as the name suggests are Korean Massive Open Online Courses; they are designed to bridge higher education with MOE lifelong learning initiatives as well as private enterprise by promoting recruitment links between students and companies in Korea (MOE, 2016). Mobile technology in this instance mirrors its use for KOCW as it is used to provide media support services for the course content.

Socialized Pedagogy: Direct Collaboration through Texting

Although rare in the larger research canon, some experimentation has taken place using instant messaging as a means of socialized collaboration for formal learning, experimentation that proves directly applicable to the focus of this chapter and its exploration of the disconnect between formal mobile learning and informal mobile learning in South Korean higher education. Kim, Lee, and Kim (2014) detail a study of 48 students in three classes from a large private Korean university enrolled in a required introductory educational technology course using mobile messaging groups, computer-based instant messaging applications, and bulletin board applications, respectively, for course communication. The findings suggest distinctions need to be drawn between teamwork (the process of learning to work together) and taskwork (the process of completing required course activities) in the use of a particular ICT or mobile application.

A further distinction is drawn by this author in the following discussion of the informal mobile learning space in South Korea, noting that socialized activity (what Kim, Lee, and Kim might refer to as teamwork) has enjoyed a long history in South Korea; university students have many years of experience in managing and engaging in their many socialized communities through mobile technology before stepping foot on a university campus. This experience, this author argues, is ultimately neglected by universities in their attempts to position mobile technology as a node in a larger smart learning system, a supplementary tool in a larger daisy chain of technologies with considerable institutional and instructional control implicit in its design. For the students themselves, however, in their informal communities, mobile technology reigns supreme.

Close-Knit Peer Relationships in Mobile Technology: The Informal

Informal mobile learning emerges from a mature, dynamic, and often retraditionalized (Yoon, 2003), mobile field of activity. It is from this mature technological South Korean context that mobile practices are retraditionalized—filtered through South Korean sociocultural practices (Yoon, 2003)—and where the tension emerges in the formal mlearning context discussed in the previous section, where pronounced levels of institutional or instructional control are at odds with or inhibit the practices emerging from this informal context.

To begin were "*eomjijok*", or thumbtribe groups, which can be a representation of early (late 1990s to late 2000s) Korean mobile culture. According to Ok (2011), these are South Koreans who have exceedingly swift texting skills that are used to communicate in an idiosyncratic code language, designed specifically to be understood by members of that group (or tribe) only. Jouhki (2008) and Shim et al. (2008) expand on the idiosyncratic practices that emerged from these groups, from political participatory culture to informal social communication to gender-based participation and social gaming practices in mobile communities. *Eomjijok*'s significance as it applies to this chapter is to illustrate the informal practices that sit aside and problematize the formal modes of communication and learning encapsulated in the smart learning policy of higher education.

There is considerable overlap between these *eomjijok* communities and the social media cultures that emerged both in the apex and the wake of *eomjijok*. Korean social networks tend to be much smaller than their UK or US counterparts and their motivations for participation (social support, some information seeking, less casual relationships) speak to a close-knit social network that reinforces peer communities (Kim, Sohn, & Choi, 2011). Social media, like mobile technology, "reconfirms young people's peer networks, which continue traditional modes of sociality and cultural identity" (Ok, 2011), a stance that Yoon's (2003) concept of retraditionalizing would affirm and qualify; retraditionalizing involves the idea that mobile technology, or indeed any new technology, generates activity that is localized to make it aligned with existing communicative practices. South Korean mobile culture presents a unique mix of the individualized filtered through socialized practice.

As of 2015, social media use has migrated to mobile messaging applications; this shift to messaging applications suggests a parallel shift to the informal practices contained therein, what Park (2013) might refer to as the reproduction and negotiation of practices and expectations, specific to the textual medium. This shift is familiar as "the original invention and popularization of the medium as well as the habitus that preceded it is still present in the memories of many participants within the new texting habitus" (Park, 2013); this shift is especially familiar as it represents a reversion to many of the practices found in *eomjijok* cultures: idiosyncratic language and emoji acting as code, blogs as diaries and platforms for mobile media, and media as community and communicative artifact. These informal practices have spanned the *eomjijok* cultures of the late 1990s to the present day. One such application is KakaoTalk, a mobile messaging application that enjoys predominate market share. KakaoTalk is the dominant social media application and environment in South Korea at 39 per cent of all social media countrywide (We Are Social, 2015).

KakaoTalk is deeply integrated into South Korea socialized practice, both informal and formal, activated through mobile technology; it is used by university students to maintain participation across informal and formal communities. It adapts many of the design features from earlier iterations of South Korean social media, particularly as related to personal messaging narrated and presented to a close-knit network (Hjorth, 2009a) and the "migratory practices across interconnected digital media spheres" (Hjorth & Kim, 2005). KakaoTalk is designed to maintain close-knit networks through digital communication that enhances connections across a range of communication channels: every sticker picture, emoticon, and filtered avatar ("cute aesthetics"); every game played collaboratively; every chat detailing the activities and observations of the day (a personal narration); and every movement between one community and another ("migratory practices across interconnected digital media spheres"). KakaoTalk inexorably and implicitly links these communities, even encourages migration between them, by presenting them simultaneously in its interface. Once linked, KakaoTalk allows for maintenance of community participation across a myriad of media channels with little effort; every emoticon builds connection across informal communities and every screenshot of a lecture slide or audio recording of a group discussion builds camaraderie in formal, academic communities.

KakaoTalk is not a node on a larger smart-learning schematic, but rather at the nexus of learning itself. This learning takes on distinct South Korean forms, undergoing a "retraditionalization" (Yoon, 2003) of South Korean practice. Yoon (2003) advances the idea that much of the idiosyncratic nature of this mobile learning is driven by uniquely Korean concepts, such as *jeong* (Korean: 정), defined as:

> [A]n expression of affective and attached relationships between people closely related to one another. *Jeong* is, on the one hand, based on an

> extended form of familism in that it emphasizes the strong attachment between close people within the network … . Once someone begins to be considered as a member of the network, he or she is treated as family by other members and, in consequence, it becomes extremely difficult to keep away from the network
>
> *Yoon, 2003*

Jeong, as such, acts as a binding agent, a local sociality as described by Yoon (2003), governing or structuring social interaction that governs both informal and formal mobile learning. Further binding agents manifest formally in the disciplinary space in intricate "senior–junior" (Korean: 선배/후배, or *seonbae/hubae*) relationships that permeate higher education; these are mentoring relationships between younger and older students in the same discipline where disciplinary practices are modeled and where a sociality is enforced. They move between mobile activity and non-digital activity routinely; they are occasionally resisted and contested in subtle acts of subversion (Gallagher, 2016).

These specific South Korean agents formalize many of the social practices emerging through mobile technology and, in turn, affect the shape of disciplinary participation, evident in the norms and importance of reciprocity:

> The members in the friendship circle are subject to the obligation to accept as well as the obligation to reciprocate.
>
> *Taylor and Harper, 2002*

> Text messages play a particularly important role in this sharing through the mobile by maintaining continuous connection … the continuous reciprocal ritual tends to strengthen the ties between members without intentional disconnection or 'chewing out', which refers to ignoring calls or messages from others.
>
> *Yoon, 2003*

Mobile technology intensifies this process by allowing for "individuals to maintain states of hyper-connection and hyper-awareness of others" (Lee, 2012). Every KakaoTalk discussion with friends, with family, with seniors, with juniors, with fellow classmates, or faculty, all sitting in close proximity on a mobile screen, are bound by these South Korean sociocultural bindings agents. To say that mobile technology is merely a node in this smart learning system is erroneous; mobile technology is critical to participation in South Korean society in general and South Korean higher education specifically by allowing for the capacity to maintain relationships along the nexus of multimembership (Wenger, 1998).

Findings: The Intersect and the Tension in Mobile

These two fields of mobile learning activity, the formal and the informal, exist in states of overlap and tension. They are not mutually exclusive fields of activity.

Rather, the tension discussed here is one of misalignment and a general misunderstanding of the role of mobile technology in the learning processes of university students. Formal mobile learning in Korean higher education, aside from limited research on socialized mobile learning using messaging applications (discussed earlier in Kim, Lee, & Kim, 2014), has yet to embrace or even fully employ practices or activities emerging from the informal mobile context as discussed in this chapter. Formal mobile learning is subsumed in this context in a larger smart learning system and, as such, moves to the periphery of technology enhanced learning (TEL)—a subset of a larger whole learning structure. While this presents material advantages—the efficient employ of abundant connectivity and ICT resources across higher education, for example—it implicitly negates the importance of mobile technology in learning in South Korea. Students want more than the fragmentary or supplementary uses of mobile technology that formal mobile learning currently provides—administrative functions, content access, and limited communicative potential (Han & Han, 2014). They want formal mobile learning to enhance their capacity to manage their multimemberships and, in turn, manage their learning in socioculturally specific ways. The supporting role afforded mobile technology in a smart learning structure (as supplementary to cyber universities, open courseware, and open learning, as discussed earlier) contrasts significantly with the privileged role afforded mobile technology by the university students using it.

Formal mobile learning in higher education is hampered as a result, evidenced by the resistance to both smart learning and mobile learning in the research (discussed in Park, Nam, & Cha, 2012). Mobile learning in South Korean higher education is, essentially, siloed from this wellspring of mobile activity evidenced by South Korean university students in their informal activities; as such, it is positioned askance when initiated formally in universities. Formal mobile learning applications developed by universities to support particular aspects of learning or disciplinary participation pale in comparison to the range of activity, learning or otherwise, being supported in commercial applications such as KakaoTalk and other domestic applications. Considering the structural shifts brought about by consolidation due to dwindling birth rates, there is opportunity for revisiting formal mobile learning, to develop pedagogies to support that mobile learning, and to do so in tandem with the informal activity made manifest in commercial mobile applications.

The Way Forward: Conclusion and Future Research

As discussed, mobile learning in South Korean higher education is bound in problematic tension between a top-down model of smart learning in the formal space and the sophisticated use of mobile technology and social media in the informal space, use that directly affects the nature and structure of formal disciplinary participation in higher education. This chapter explored this tension from the formal as a response to university consolidation and reorganization, and from the informal as a product of a mature and sophisticated mobile system of activity.

Further research is required to explore pedagogies that bridge formal and informal mobile learning, particularly in specific sociocultural contexts. This author believes that sociocultural practice dictates much of the success of any pedagogical approach, ICT-based or otherwise, and an effective bridge between formal and informal mobile learning would require sociocultural fidelity to lived South Korean practice. Further research is required to explore the nature of proximity in mobile learning and how that influences the mobile learner's understanding of community boundaries. In mobile messaging applications where much of this learning activity is taking place, text threads that span formal learning (faculty, senior–juniors, classmates) sit astride those for informal learning (family, friends, etc.). Practices are shared and presumably boundaries blur. Further research is needed to explore the role of this proximity in shaping participation in higher education.

References

Choi, H., Woo, Y., & Jung, H. (2013). Students' perception of smart learning in distance higher education. *The Journal of the Korea Contents Association, 13*(10): 584–593. Retrieved from http://www.dbpia.co.kr/Article/NODE02369561

Chosun.com (2016). Ministry of education releases report on condition of universities. Limits of government funds to poorly ranked universities to continue; some universities (Daegu Foreign Language University and Seonam University etc.) at risk of closing. Retrieved from http://news.chosun.com/site/data/html_dir/2016/09/05/2016090502325.html

Gallagher, M. (2016). Charting trajectories on the peripheries of community practice: Mobile learning for the humanities in South Korea. (Unpublished doctoral thesis). University College London, Department of Culture, Communication and Media, London. Retrieved from http://michaelseangallagher.org/wp-content/uploads/2013/08/Gallagher_Thesis.pdf

Han, I., & Han, S. (2014). Adoption of the mobile campus in a cyber university. *The International Review of Research in Open and Distributed Learning, 15*(6). Retrieved from http://www.irrodl.org/index.php/irrodl/article/view/1950/3132

Hjorth, L. (2008). Being real in the mobile reel: A case study on convergent mobile media as domesticated new media in Seoul, South Korea. *Convergence: The International Journal of Research into New Media Technologies, 14*(1), 91–104.

Hjorth, L. (2009a). Gifts on presence: A case study of South Korean virtual community, Cyworld's mini-hompy. In G. Goggin & M. McLelland (Eds.), *Internationalising the Internet*. London: Routledge.

Hjorth, L. (2009b). *Mobile media in the Asia Pacific: gender and the art of being mobile*. London: Taylor & Francis.

Hjorth, L., & Kim, H. (2005). Being there and being here: Gendered customizing of 3G mobile practices – Through a case study in Seoul. *Convergence: The International Journal of Research into New Media Technologies, 11*(2) 49–55.

Ho Cheong, J., & Park, M. C. (2005). Mobile internet acceptance in Korea. *Internet Research, 15*(2), 125–140.

ITU ICT (2015). Internet connectivity and mobile subscription rates by country. Retrieved from https://www.itu.int/en/ITU-D/Statistics/Documents/facts/ICTFactsFigures2015.pdf

Jouhki, J. (2008). The emotional technology of tomorrow: the visual and textual rhetoric of promoting a ubiquitous technology society in Korea. In *Proceedings of the IADIS Multi Conference on Computer Science and Information Systems*, Amsterdam, Netherlands, July 2008.

KERIS. (2015). White paper on ICT in Education Korea. Retrieved from http://english.keris.or.kr/es_ac/es_ac_210.jsp

Kim, H., Lee, M., & Kim, M. (2014). Effects of mobile instant messaging on collaborative learning processes and outcomes: The case of South Korea. *Educational Technology & Society*, *17*(2), 31–42.

Kim, Y., Sohn, D., & Choi, S. M. (2011). Cultural difference in motivations for using social network sites: A comparative study of American and Korean college students. *Computers in Human Behavior, 27*(1), 365-372.

Korean Council for University Education. (2016). Closure and change through interdisciplinary university status. Retrieved from http://www.academyinfo.go.kr/UIPISA/html/egovframework/uipisa/uipnh/unt/ipsrch/Pop_mergeschool_tab02.html

Korean Statistical Information Service. (2016). Retrieved from http://kosis.kr/eng/

Kwon, H. I., Kim, D. J., Ryu, G. J., Kang, J. H., Park, J. S., & Joo, H.Y. (2013). The framework of the smart learning infrastructure in South Korea-Focus on Agriculture Education system. In *Proceedings of Free and Open Source Software Conference (FOSSC-13)*, Muscat, Oman, February 18–19, 2013.

Lee, D. H. (2012). In bed with the iPhone: The iPhone and hypersociality in Korea. In I. Richardson, L. Hjorth, & J. Burgess (Eds.), *Studying mobile media: Cultural technologies, mobile communication, and the iPhone*. Routledge Research in Cultural and Media Studies. London: Taylor and Francis.

MEST. (2009). Higher Education Act of South Korea. (Referenced in Presidential Commission on Education Reform. (1997). Education reform for the 21st century. Seoul, South Korea).

Ministry of Education. (2016). Main policy report for higher education. Retrieved from http://english.moe.go.kr/enMain.do

Ministry of Education, Science and Technology, Republic of South Korea, KERIS. (2012). Adapting education to the information age. Retrieved from http://english.keris.or.kr/es_ac/es_ac_210.jsp

Ministry of Government Legislation. (2009). Higher Education Act. Retrieved from http://buff.ly/2cm1VfS

OECD. (2014). Fertility rates. Retrieved from https://data.oecd.org/pop/fertility-rates.htm

Ok, H.Y. (2011). New media practices in South Korea. *International Journal of Communication*, *5*, 320–348.

Park, D. K. (2013). Texting the new habitus: (Re) producing and negotiating practices and expectations of the texting medium. Retrieved from http://digitalcommons.bard.edu/cgi/viewcontent.cgi?article=1232&context=senproj_s2013

Park, S.Y., Nam, M. W., & Cha, S. B. (2012). University students' behavioral intention to use mobile learning: Evaluating the technology acceptance model. *British Journal of Educational Technology*, *43*(4), 592–605.

Ryan, J., & Louie, K. (2007). False dichotomy? 'Western' and 'Confucian' concepts of scholarship and learning. *Educational Philosophy and Theory*, *39*(4), 404–417.

Shim, T., Kim, M., & Martin, J. (2008). *Changing Korea: Understanding culture and communication*. New York: Peter Lang.

Taylor, A. S., & Harper, R. (2002). Age-old practices in the 'new world': A study of gift-giving between teenage mobile phone users. In *Proceedings of the SIGCHI conference on Human factors in computing systems* (pp. 439–446), Minneapolis, MN, April 20–25, 2002.

We Are Social. (2015). Digital, social, and mobile worldwide 2015. Retrieved from http://wearesocial.com/uk/special-reports/digital-in-apac-2016

Wenger, E. (1998). *Communities of practice: Learning, meaning, and identity*. Cambridge, UK: Cambridge University Press.

Yoon, K. (2003). Retraditionalizing the mobile young people's sociality and mobile phone use in Seoul, South Korea. *European Journal of Cultural Studies*, *6*(3), 327–343.

13

EMANCIPATING EDUCATION IN CRAFT APPRENTICESHIPS USING MOBILE LEARNING

Jim G. Ffrench and Barry J. Ryan

Introduction

Craft apprenticeships are time honoured and have been used on a widespread basis for centuries going back to medieval guilds. Apprenticeships are a structured training system, which encompass challenging standards and high expectations of proficiency, leading to certification in a particular craft (Hayes, 2012). Apprenticeship learning consists of both on-the-job, one-on-one training between a mentor and an apprentice, and off-the-job, classroom and laboratory/workshop instruction on the principles of the craft.

Theoretical Background

The Demotivating 'Mathematics Problem' in Apprenticeship Learning

Apprenticeship learning is fundamentally based on interpreting problems, understanding tasks and then completing predefined duties in challenging circumstances (Collins, Brown, & Newman, 1989). In Ireland, it is expected that apprentices have acquired a minimum educational qualification of a pass in five subjects in the Irish Junior Certificate examination (or equivalent); however, their mathematics qualification level is not specified. As such, a significant proportion of apprentices experience difficulties with mathematics, plotting charts and basic mathematical process skills (McMahon, 2010). In general, the mathematical preparedness of third-level Irish students has declined over the past decade (Gill, Mac an Bhaird, & Ni Fhlionn, 2010); this has been labelled by commentators as the "mathematics problem" (Savage, 2003).

Constructing Motivation: The Attention Relevance Confidence Satisfaction (ARCS) Model

"The motivation level of learners [is] the most important factor in successful instruction" (Dick & Carey, 1996, p. 92).

Motivated students actively seek to acquire the knowledge or skill that an academic activity is designed to develop. Motivated learning is a cyclical process, and for a learning experience to be successful students must be properly motivated; students who are successful in learning will be more motivated to continue doing so in the future (Hodges, 2004). Keller (1987a) suggests that the ARCS motivation model is a pragmatic approach to incorporating motivational strategies into instructional design (see Figure 13.1). The model is based on its acronym (**a**ttention, **r**elevance, **c**onfidence, and **s**atisfaction), and is based on motivational concepts and a problem-solving approach to design which is appropriate for, and aligned to, apprenticeship learning (Keller & Suzuki, 2004, p. 229).

The ARCS model is primarily grounded upon the Expectancy-Value theory of Tolman (1932) and Lewin (1935) in which motivation is defined as a function of the expectancies and the values of the learner. The model is practical and focuses on matching motives to the unique affective characteristics of learners. Keller (1987a) defined motivation as the value attributed to attainment of a goal and the expectancy that the goal can be accomplished given specific

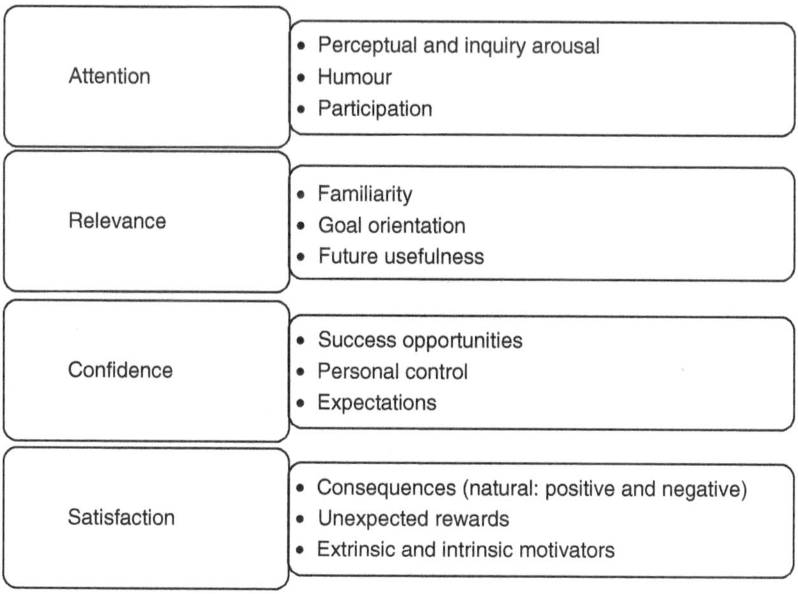

FIGURE 13.1 Overview of the ARCS model of motivation in learning. Each of the four stages of the model is supported by a strategic element approach that permits instruction to be motivational (Keller, 1987a; 1987b; 1987c).

circumstances. Students in apprenticeship classes are primarily interested in relevance and immediate application of learning and this "just-in-time" requirement for knowledge, understanding and application feeds into the integrated nature of mobile learning (mlearning) where learning takes place irrespective of context and time (Traxler, 2009).

Motivated Mobile Learners: A Novel and Appropriate Apprenticeship Pedagogy

Jones and co-workers (2006) identified six reasons why mlearning can be motivating (see Figure 13.2) and this supports the application of mlearning to apprenticeship teaching and learning.

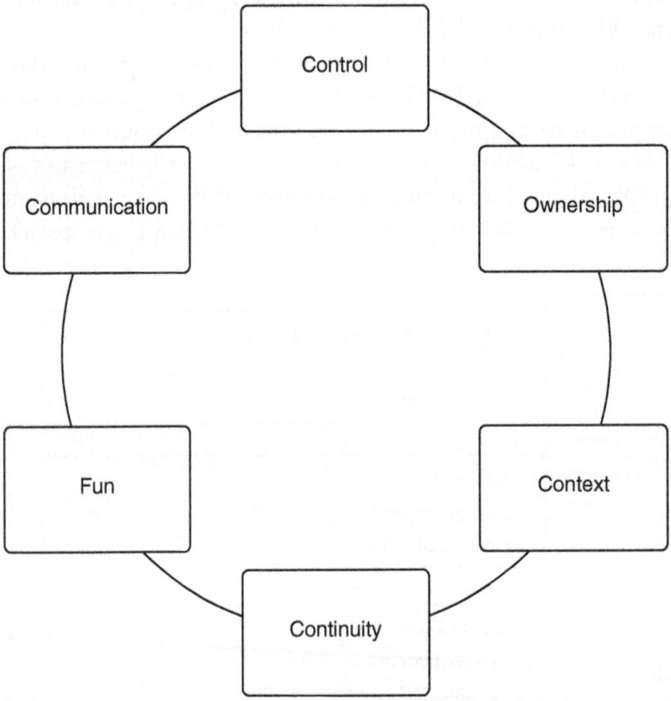

FIGURE 13.2 Jones and colleagues' (2006) six interrelated categories of motivation by mlearning. These suggest that (1) students have *control over their own goals*; (2) mobile devices afford students an important awareness of control and *ownership*; (3) mobile devices empower students to find and share resources in the *context* where they are needed; (4) mobile devices can provide *continuity between different contexts* because of their portability; (5) mobile devices are "*fun*" devices and may stimulate student motivation to learn; and (6) mobile devices enable *communication* between students on collaborative learning activities which can be motivating.

Apprenticeship learning is a specialized sub-set of further and higher education; provision needs to be flexible, dynamic and pragmatic to allow learners to seamlessly engage with on-site and in-class learning. Achieving this through traditional pedagogies may not be the most appropriate approach. Park (2011) suggests that mlearning enables individual learners to decide where and when to learn, and to personalize their learning environments. Supporting this concept, Naismith and co-workers (2004) defined six theory-based classifications of learning promoted by mlearning: namely, the *behaviourist-type*, *constructivist*, *situated*, *informal* and *support of learning* and *resources*. These definitions can be mapped onto the various modes of apprenticeship learning; for example, a behaviourist mlearning activity can give a quick feedback or reinforcement element, and informal learning opportunities allow the learner to use the mobile device as an expedient source of information on the go (Traxler, 2009). In terms of student motivation, Shellnut (1996) suggests that Keller's motivation design concepts "apply equally well to the three major learning approaches used in the field today – behaviourism, cognitivism, and constructivism".

Mlearning affords substantial opportunities for genuinely supporting differentiated, autonomous and individualized learning through mobile devices (Pachler, Bachmair, & Cook, 2010). Mlearning, combined with judicious instructional design guided by the ARCS motivational model, was chosen as the basis for this action research into a novel and appropriate apprenticeship pedagogy as it can encourage learning in a real-world context, and help bridge classroom, workshop, commuting and home environments (Gil-Rodríguez & Rebaque-Rivas, 2010). In this study, mlearning offered a real possibility to scaffold students' learning by affording them the opportunity to create their own "personal learning narratives", through an extended process of accessing and assimilating new knowledge and organizing situated activities (Sharples, 2000). These personal learning narratives, combined with situated learning, offered the learner the opportunity of highly contextualized and realistic events within which learning can present itself (Low & O'Connell, 2006).

The research outlined here investigated the efficacy of bespoke mlearning resources in order to determine the effect the mlearning resources had in motivating the students in learning a mathematics-related part of their syllabus, specifically the vapour-compression refrigeration cycle. This research was carried out in an Irish Institute of Technology that caters for all levels of tertiary education, ranging from Level 6 (including apprenticeship) to Level 10 (PhD), based on the Irish Quality and Qualifications National Framework. The school supporting the research has over 100 years of experience catering for apprenticeship education and training, with crafts ranging from plumbing through cabinet making, and more recently refrigeration and air conditioning. The findings from this research, in the specific context of this research and also more generally in apprenticeship teaching and learning, will be discussed, conclusions drawn and recommendations for practice will be offered.

Study Design

Whitehead and McNiff's (2006) Action Research Model formed the basis of this research and was comprised of four stages: Planning, Acting, Observing and Reflecting (see Figure 13.3). The research was itself bounded by the research question:

Can the introduction of mobile learning resources in the final year of a four-year apprenticeship in refrigeration enhance student perceived motivation to learn key threshold disciplinary concepts?

Planning

A situation analysis underpinned the planning phase and comprised of interviews with academic contemporaries (n = 2) with extensive experience in teaching and training refrigeration apprentices, to establish the nature and extent of the apprentices' issues with the vapor-compression refrigeration cycle. These two contemporaries also acted as mentors throughout the planning phase, piloting and reviewing the resources and providing valuable feedback.

A questionnaire-based student-needs analysis was executed to determine how the apprentices (n = 12) could be assisted in learning the mathematical process skills required. A key aspect of the needs analysis aimed to identify, and match, the learning needs of the students to the development of the proposed resources.

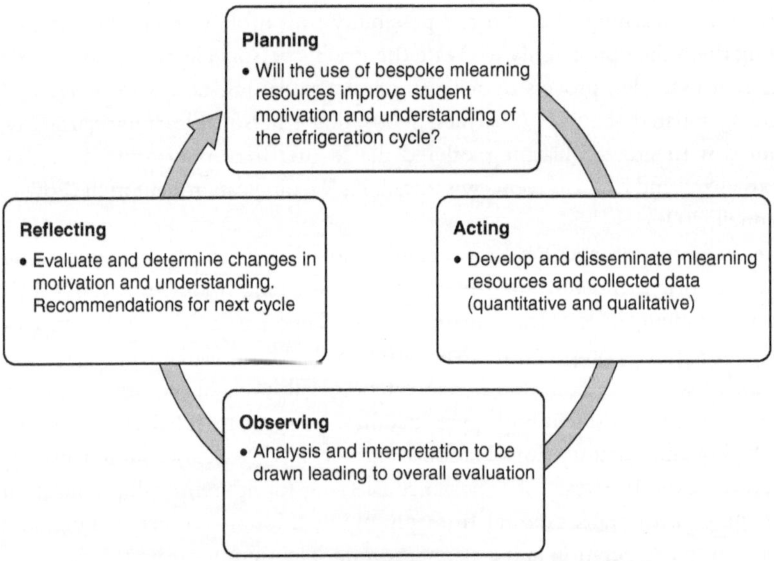

FIGURE 13.3 A schematic overview of an adapted Action Research Cycle (adapted from Whitehead & McNiff, 2006); containing four steps: *Planning*, *Acting*, *Observing* and *Reflecting*.

Having analyzed the data from the draft questionnaire, strict learning objectives for the resources were determined and a comprehensive storyboard, adhering to Gagné and Dick's Events of Instruction (1983), was constructed to develop the structure and layout of the resources leading to the *Acting* phase of the project.

Acting

In designing the resources for this research, many strategies and tools were employed to satisfy the categories of Keller's ARCS model and thereby motivate the students to learn.

Articulate Storyline® was chosen as the eAuthoring tool to allow the resources created to be accessible on a mobile platform. The tool allowed the resources to be published to Adobe Flash, HTML5 and a version optimized for iPad®. The resources were hosted on www.articulate-online.com to take advantage of the student analytics provided. The resources comprised short interactive learning resources, focusing on key, troublesome points of the curriculum. Students were encouraged to use the resources as a just-in-time learning approach and also as part of a synoptic revision strategy. Initially the resources were made available as online smartphone compatible objects; however, they were also supplied as generic mlearning resources (i.e. laptop and smartphone compatible). The resources were available for the majority of the course and the students could access the resources as a preparation tool before lectures, during relevant laboratory classes and as a revision tool after lectures.

Observing

A mixed methods approach was taken in this study, which employed a range of data collection techniques (Bryman, 2012; Denscombe, 2010; Klassen, Creswell, Plano Clark, Smith, & Meissner, 2012). To ensure the data validity, triangulation, using three different methods involving five techniques, was adapted, i.e. a questionnaire for breadth, focus group interviews for depth and student observations, analytics and assessment results for corroboration (Johnson & Onweugbuzie, 2004).

Data Collection

All data were collected in accordance with the host institutes Ethics Committee Guidelines and the British Educational Research Association (BERA) guidelines (2011), and Hammersley and Traianou (2012). Volunteered consent was recorded from all participants.

Qualitative data was collected using a questionnaire and was administered to the cohort of students at the beginning of their ten-week course. In week eight of the course, further data was collected using three focus group interviews, each with four students. The focus groups' interview questions were influenced by the themes that emerged from the data received in the participant questionnaires.

The semi-structured interviews were conducted by an independent colleague and were audio recorded, which permitted emerging themes to be investigated further within the interviews (Anderson, 1996; DiCicco-Bloom & Crabtree, 2006; Tracy & Carmichael, 2010). Qualitative data were analyzed manually by means of thematic content analysis, as described by Stewart, Shamdasani and Rook (2006). Data here included discussion fora and interviews. Additionally, a reflective action research journal was kept by the researcher, wherein additional insights were recorded on a weekly basis by direct observation of student interactions with the mlearning resources and verbal exchanges with the students. This data informed additional questions for the focus group interviews and allowed for analysis and interpretation, inferences to be drawn and further evaluation and determination of necessary changes from an insider researcher.

Quantitative data collected through the initial questionnaire and focus group interviews were transferred into Excel spreadsheets to permit in-depth analysis and visualization. Additional data in this set included student usage analytic data provided by the host site; these data were recorded and interrogated. Data included individual resource view counts, completion rates, quizzes progress, results and general consumption trends. These metrics were reported individually for student-based intervention and in aggregate for resource-based analysis. The student analytics were recorded throughout the time the resources were available. The students' summative assessment results for the respective subject matter question were also recorded. Furthermore, student assessment results for this specific section of the curriculum were also statistically analyzed in comparison to the results of thirty previous student cohorts, assessed between 1996 and 2012, during which time the curriculum remained consistent, for corroboration of the potential of the resources.

Overall, an inductive approach was taken to the quantitative and qualitative data analysis. The combined data were analyzed to establish if there was a verifiable motivational change to the students' predispositions to the subject matter and whether mlearning had contributed to this change.

Study Design: Summary Reflection

The following section will discuss the findings of the study, cognizant that the perceptual data reflects the views of a small sample size (n = 12 students) and can neither be generalized to refrigeration apprentice students nor apprentice students as a whole. Likewise, the study's results may not be transferable to students in other disciplines. Furthermore, the researcher had no control over the type of device used by the students to access the mlearning resources; and as such the mlearning devices discussed here will include both smartphones and laptops. In brief; three main findings emerged from the in-depth analysis of the data and these will be explored in a reverse hierarchical order.

Barriers to Access

The main finding of the research was that barriers were associated with accessing the mlearning resources. As five of the students did not have a mobile device capable of accessing the mlearning resources, they used their laptop computers to access a copy of the combined resources online. However, after the initial novelty of using their mobile devices, the other seven students primarily chose to use their laptop computers to access the online copy of the resources, in preference to the mlearning platform resources. This echoes Stockwell's (2008) study, which revealed that students used the personal computer in preference to the mobile device in the majority of cases. During the analysis of the action research data described here, four barriers emerged as to why the students did not use the mlearning resources as much as their questionnaire responses indicated.

Technology

Most higher education learning environments include technology to support teaching and learning, be they self-directed or instructor-facilitated (Keller, 2008b). However, personal technical issues can make the adaptation of mlearning demanding and this can impair student inclination to accept mlearning (Wang, Wu, & Wang, 2008). All of the students identified physical aspects of mobile technologies, such as small screen size, processing speed, excessive download time and operating system/browser compatibility as issues that influenced their reluctance to use the mlearning resources. One student had access to two smartphones (Blackberry and Sony), neither of which was able to access the mlearning resources due to platform compatibility issues. Six of the students (50 per cent) had mobile devices with no internet capability, while eight of the students (67 per cent) cited that small screen size was an issue. One student stated that the "smartphone graphics were just about viewable" and that "pinch and zoom only worked on some parts of the resources" (rec.3, 26-03-2013). These technical barriers concurred with previous findings in the literature (Lawrence, Bachfischer, Dyson, & Litchfield, 2008; Williams, 2009).

Cost

The literature suggests that the cost of internet access is one of the primary barriers to students utilizing mlearning resources (Aderinoye, Ojokheta, & Olojede, 2007; Donaldson, 2011) and while network data download prices have fallen in some countries they still remain comparatively high in Ireland. In this research, 67 per cent of the students stated a reluctance to access the mlearning resources due to the cost of using 3G/4G for downloading, with one stating:

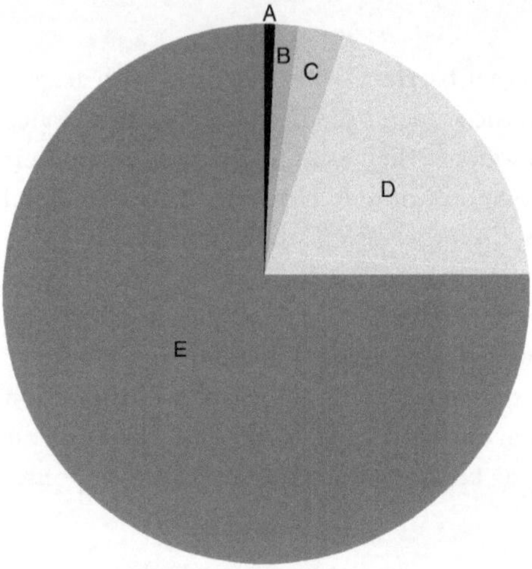

FIGURE 13.4 A comparative analysis of student activity, based on access times, between mobile device access versus laptop or personal computer access. Data was collected via student usage analytics provided by the host site. A = mlearning Module 2 (Theory), B = mlearning Module 3 (Examples), C = mlearning Module 4 (Sample Calculations), D = mlearning Modules 5 & 6 (Quizzes), E = Combined resources for laptop use.

"I would not use the resource on 3G, only on Wi-Fi" (rec.3, 26-03-2013). Three students (25 per cent) stated that they were not prepared to use their limited data download or pay network charges for mlearning (rec.2, 26-03-2013). All the students said they preferred to use their laptop computers, with two students (17 per cent) saying they were using laptops for project work anyway (rec.3, 26-03-2013). The students accessed the online copy resource using their laptop computers over 90 per cent of the time. Figure 13.4 provides a summary of the extent the individual mlearning resources were accessed (9 per cent) compared to the combined online copy resource for the laptop computers (91 per cent).

Limited Access to Broadband Wireless Internet

Poor wireless internet access has been cited as prohibiting ready access to mlearning resources (Lawrence, Bachfischer, Dyson, & Litchfield, 2008; Patrichi, 2010; Zhu, Hu, & Guo, 2012). All of the students in this research stated they found a lack of Wi-Fi in most of the areas in which they frequented, including many parts of their host institute. Where Wi-Fi was accessible, the slow download speeds deterred repeat usage. In addition, four of the students (33 per cent) stated the resources were too slow to run using Wi-Fi available to them (rec.3, 26-03-2013).

Time

A number of the students, although making positive comments about the concept of mlearning, viewed mlearning more as an additional time-consuming and potentially costly task than as a learning tool to help their motivation and understanding of a conceptually difficult part of their syllabus. The students had 35 hours of class contact per week, and upon investigation it was also found that most of the students commuted by driving to and from the institute, allowing little opportunity to use commuting time for mlearning. In addition, a student who used public transport stated, "The journeys were too short to be of any use" (rec.3, 26-03-2013). One student stated the reasons he did not access the resources more were "too little time" and "the resources were only accessed from my home at weekends … [because of] … poor network access in the accommodation" (rec.3, 26-03-2013). Another student referred to the "time required for study and the project workload" as causal factors (rec.1, 26-03-2013).

Positive Conceptual Attitude: Online Support

Despite the apparent large number of barriers to adoption, the students in this study welcomed the mlearning resources, in concept, if not in practice. Prior to their release the students foresaw themselves, on average, accessing the resources four to six times a week. However, the analytics showed, on average, the students used the resources approximately twice a week, but by using their laptop computers more than 90 per cent of the time. All of the students found the resources easy to navigate, both the mlearning and online adapted versions, and felt that the resource graphics enhanced their learning experience. All of the student cohort completed the quizzes and stated they found the scoring useful feedback. One student stated that the quiz scoring "gave you a reason to do it again to get 100 per cent" (rec.2, 26-03-2013). All of the students believed the resources' content was relevant to their learning and assessment requirements, and referred to the visual and sequential design being useful for revision. However, three students (25 per cent) suggested that if mlearning resources were designed around a course project or an alternative area deemed more relevant to industry usage, they would be utilized more (rec.1, 26-03-2013). This difficulty in motivating apprentices to learn material detached, in their minds, from skills more directly connected to the demands of their work in industry, concurs with the findings of Sagan and co-workers (2005) and in motivating mature students illustrated by Casey and colleagues (2006).

Of particular significance in terms of this study was the indication by all of the students that they would prefer if more resources were available online, accessible using a smartphone, tablet or laptop computer. The students stated a desire to have these resources cover a greater area of their syllabus and assessments with additional interactive module notes. One student suggested "more quizzes" (rec.2, 26-03-2013), and another suggested quizzes with "more random[ized] questions" (rec.3, 26-03-2013).

Increased Student Motivation

Analysis of the post mlearning intervention, through Keller's motivation lens, indicated that student motivation increased after using the resources, be they smartphone or laptop based. Keller (2008b, p. 175) stated "learning concerns learner motivation"; therefore students, who are intrinsically motivated are more likely to engage in the task and thereby gain understanding (for a summary of this, see Table 13.1). Seven of the 12 students (58 per cent) reported, in the introductory questionnaire, they had experienced learning difficulties in the area of plotting refrigeration cycle charts, determining values and using formulae. In addition, nine students (75 per cent) responded negatively when asked if they understood how to plot refrigeration cycles. Later, in the focus group interviews, all of the students stated they found the resources helpful in motivating them to learn more about the refrigeration cycle, with one student stating: "the resources and quizzes were useful combined with the class notes" (rec.1, 26-03-2013). All of the students believed they were more likely to attempt the relevant questions in their assessments after having access to the resources. One student stated, "I had not attempted the questions on this area of the syllabus in all previous assessments and I am 100 per cent more confident now" (rec.1, 26-03-2013). It is worth noting in all refrigeration apprentice written assessments there is no choice; all questions are mandatory. None of the students felt the mlearning resources impeded their learning in any way. Rather, one stated, "both the resources and classes were good" (rec.1, 26-03-2013). When the students were asked if the resources assisted in their understanding of how to plot refrigeration cycles they all responded with an emphatic "yes", referring to the resources being good for revision, allowing them to repeat the more difficult sections and self-assessing as often as required. One student stated, "you could spend as much time as you wanted at it [the resources], moving at your own pace" (rec.1, 26-03-2013).

The average pass rate for the research cohort calculated from the student summative assessment results, for the primary relevant question relating to the mlearning resources, was 75 per cent (see Figure 13.5). In comparing this figure (75 per cent) with that of the average pass rate of 30 previous classes (56 per cent) for the relevant question, assessed between 1996 and 2012, the correlation was positive, with 25 of the 30 classes (83 per cent) having a lower pass result. This indicates the resources were beneficial, and concurs with previous studies, which revealed that achievement levels increase when students utilize mobile technologies (Williams & Bearman, 2008; Wyatt, Krauskopf, Gaylord, Ward, Huffstutler-Hawkins, & Goodwin, 2010).

In summary, these results, along with the students' responses in the focus group interviews, were a strong indication that the mlearning resources had a positive influence on student motivation to learn and helped to encourage the students to gain an understanding of the topic covered. Similar positive attitudes toward mobile technologies were found in studies by Brand and co-workers (2011). Of interest was the fact that all the students provided positive comments about the

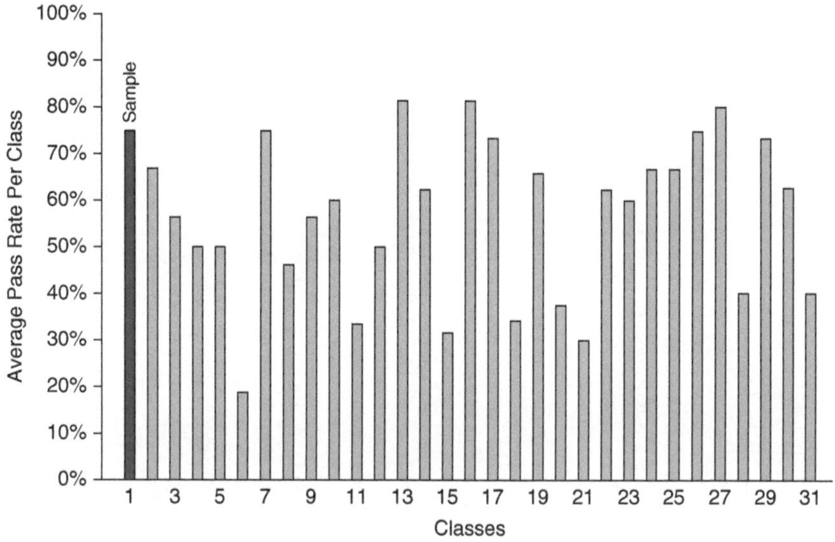

FIGURE 13.5 A comparative analysis of the average student (n = 12) pass rate for the primary relevant question for the cohort that experienced the mlearning intervention in comparison to the previous 30 classes that experienced traditional apprenticeship pedagogies. The typical class size for the traditional apprenticeship pedagogies cohort was 16.

concept of mlearning, even those students who did not use mobile devices, and this chimes with Stockwell's (2008) study.

Conclusions, Implications and Recommendations

"Data from qualitative research are typically suggestive, rarely if ever conclusive" (Sofaer, 2002, p. 334). That said, the results of this study indicate that using mlearning resources, as a means for incorporating the motivational first principles into instruction, went some way to bridge the gap in student motivation towards learning the relevant science and mathematical process skills required for the specific area addressed by this research. However, this study found the cohort of students, while positive about mlearning as a concept, did not have a strong willingness to adopt it.

Mobile devices give a greater sense of freedom of time and place to learn, but "freedom is about experimentation and making choices" (Stockwell, 2010, p. 108). While the students were encouraged to access the resources using a mobile device, a copy was made available online also. These apprentice students, often time-poor for study, wary of the personal network costs of mlearning, experiencing access difficulties due to a lack of Wi-Fi areas and download speeds, and the difficulties incurred with using small screen size, deferred to the perceived easier

TABLE 13.1 A Summary of the ARCS Model of Motivation Applied to Learning, Detailing the Strategic Element of Motivation, with Exemplars from This Study Linked to Tangible Outputs. Relevant References Are also Cited.

ARCS Stage	*Strategic Element of Motivation*	*Exemplar Method*	*Tangible Output*	*References*
Attention	Perceptual/inquiry arousal	Questionnaire to gain insight into the students' opinions and predispositions to the area of the refrigeration cycle, mobile device usage and mlearning.	Trigger students' attention and curiosity.	Keller, J. (1987b)
	Participation and persistence	Appropriate humour and quizzes were embedded.	Data analytics of student interaction.	
Relevance	Content familiarity	Authentic content, appropriate assessment.	Students as self-determined, active agents of knowledge creation.	Duffy, Lowyck, & Jonassen (1993)
	Goal orientation	Worked examples of real life problems.		
Confidence	Success opportunities	Appropriate in-resource quizzes and assessments.	Students achieve required competencies to pass their summative assessments for industry competence and affiliation.	Keller (1987a)
	Personal control/self-paced	A straightforward, consistent and easily understood navigation system.		
Satisfaction	Designed for all learner types	Synced audio and visual components with embedded interactive learning events.	Students examined their developing understanding and identified failures in assumed understanding.	Brown, Joanna, & Race (1999)
	Instant feedback	Instantaneous feedback provided.		
	Opportunities to repeat/reuse	Students could return to any element of the resource.	Students had opportunity to apply what they had learned in course projects.	
	Positive/negative consequences		Student acceptance of the *equity* of result with effort.	
Maintaining	Employ volitional or self-regulatory strategies	Bite-sized resources allowing for parsimonious information processing.	Maintaining students' goal-oriented behavior and overcoming discouragement and attrition.	Keller (2008a)

option of elearning. In practice, they preferred to use their laptop computers for accessing the online copy of the resources, as they were regularly using their laptop computers for course project work anyway.

Implications for Future Practice and Research

Through mlearning, there is the prospective of providing a rich learning environment for students, but there are still issues that must be considered before it can reach its full potential. The disconnect between technology's promises and students' expectations can still be frustratingly wide. Although mlearning has evolved over the last decade, becoming universal, easier and cheaper, Traxler and Kukulska-Hulme (2015, p. 2) suggest there is "considerable granularity when comparing individual countries and specific economic classes". The researcher's belief that mlearning would be a valuable contribution in the teaching of apprentices was based on his social and cultural influences and these may have differed from the students' influences (Crompton, 2016). In addition, Crompton (p. 36) identifies three types of resources as "essential for effective technology integration": training, access to mobile technologies and technical support. While the researcher undertook available training, there was a large percentage of students who were unable to access the resources. Furthermore, the lack of technical support available to the researcher on the specific issues surrounding the eAuthoring tool and the online platform impinged on the initial creation and the hosting of the resources. From an institutional perspective while the use of technology is being embraced in program delivery, pedagogical practice has not kept pace (Murthy, Iyer, & Warriem, 2015). In contrast to Wu and Chao's (2008) findings, mlearning is unlikely to be the platform of choice in apprenticeship learning until these issues are resolved.

From a broader societal viewpoint, the Irish economic recovery goes hand-in-hand with a high demand for new apprentices, particularly in the construction industry. The Irish Construction Industry Federation (CIF) has recently launched a new website (www.apprentices.ie) for apprentices seeking opportunities to meet this need. This increase in apprentice numbers brings opportunity to improve the efficacy of their off-the-job third-level education.

The study has shown mlearning has the potential to provide a rich learning environment for students, offering "true self-paced opportunities for learning, anytime and anywhere" (Haag, 2011, p. 12), but until ongoing issues are addressed, this will not be fully realized. In considering how mlearning could be used to provide students with better opportunities and enhanced learning outcomes, Stockwell (2010, p. 108) states "successful use of mobile technologies relies on keeping up with our changing learners, and continuing to give them opportunities to experiment and discover". This study has explicitly shown that mlearning has its place in apprenticeship learning; however, finding its most proficient application in the apprenticeship area will require further research. The findings of this research are of relevance to academic staff involved in apprenticeships and who are considering the introduction of mlearning resources as part of a blended learning program and working within the national curriculum constraints. The study's findings should encourage engagement with this potential emancipatory approach to an age-old model of learning.

References

Aderinoye, R., Ojokheta, K., & Olojede, A. (2007). Integrating mobile learning into nomadic education programmes in Nigeria: Issues and perspectives. *International Review of Research in Open and Distance Learning, 8* (2).

Anderson, P. (1996). Ethics, institutional review boards, and the involvement of human participants in composition research. In P. Mortensen & G. Kirsch (Eds.), *Ethics & representation in qualitative studies of literature* (pp. 260–286). Urbana, IL: National council of teachers of English.

BERA. (2011). Ethical guidelines for educational research. Retrieved from https://www.bera.ac.uk/researchers-resources/publications/ethical-guidelines-for-educational-research-2011

Brand, J., Kinash, S., Mathew, T., & Kordyban, R. (2011). iWant does not equal iWill: Correlates of mobile learning with iPads, e-textbooks, BlackBoard Mobile Learn and a blended learning experience. In G. Williams, P. Statham, N. Brown, & B. Cleland (Eds.), *Changing Demands, Changing Directions* (Vol. *1*) (pp. 168–178). Hobart, Tasmania: Proceedings Ascilite.

Brown, S., Bull, J., & Race, P. (1999). *Computer-assisted assessment in higher education*. London: Routledge.

Bryman, A. (2012). *Social research methods*. Oxford, UK: Oxford University Press.

Casey, H., Jupp, T., Grief, S., Hodge, R., Ivanic, R., Lopez, D., … McNeil, B. (2006). *'You Wouldn't Expect a Maths Teacher to Teach Plastering'. Embedding, Literacy, Language and Numeracy in Post-16 Vocational Programmes – the Impact on Learning and Achievement*. London, UK: National Research and Development Centre for Adult Numeracy and Literacy.

Collins, A., Brown, J., & Newman, S. (1989). Cognitive apprenticeship: Teaching the crafts of reading, writing, and mathematics. In L. Resnick (Ed.), *Knowing, learning, and instruction: Essays in honor of Robert Glaser* (pp. 453–494). New York: Lawrence Erlbaum Associates.

Crompton, H. (2016). Moving toward a mobile learning landscape: Presenting a mlearning integration framework. *International Mobile Learning Festival* (pp. 27–28).

Denscombe, M. (2010). *The good research guide: For small scale social research* (4th ed.). Buckingham, UK: Open University Press.

DiCicco-Bloom, B., & Crabtree, B. (2006). The qualitative research interview. *Medical Education, 40*(4), 314–321.

Dick, W., & Carey, L. (1996). *The systematic design of instruction* (4th ed.). New York: Longman.

Donaldson, R. (2011). *Student acceptance of mobile learning. Electronic theses, treatises and dissertations. Paper 716*. Florida State University. The Graduate School at DigiNole Commons.

Duffy, T., Lowyck, J., & Jonassen, D. (1993). *Designing environments for constructivist learning*. New York: Springer-Verlag.

Gagné, R., & Dick, W. (1983). Instructional Psychology. In M. Rosenzweig, & I. Porter (Eds.), *Annual Review of Psychology* (Vol. *34)* (pp. 261–295). Palo Alto, CA: Annual Reviews.

Gill, O., Mac an Bhaird, C., & Ni Fhlionn, E. (2010). *The Origins, Development and Evaluation of Mathematics Support Services*. Dublin, Ireland: Irish Mathematical Society Bulletin, 66.

Gil-Rodríguez, E., & Rebaque-Rivas, P. (2010). Mobile learning and commuting: contextual interview and design of mobile scenarios. In *USAB'10 Proceedings of the 6th international conference on HCI in work and learning, life and leisure: workgroup human-computer interaction and usability engineering* (pp. 266–277). Berlin Heidelberg: Springer.

Haag, J. (2011). From elearning to mlearning: The effectiveness of mobile course delivery. In *The Interservice/Industry Training, Simulation & Education Conference (I/ITSEC). 2011*, pp. 1–13. Alexandria, VA: National Training Systems Association.

Hammersley, M., & Traianou, A. (2012). *Ethics and educational research*. London: British Educational Research Association on-line resource.

Hayes, J. (2012). *Apprenticeships. Department for Business, Innovation and Skills and the Department for Education*. Retrieved from www.parliament.uk: http://www.publications.parliament.uk/pa/cm201012/cmselect/cmbis/writev/1843/app001.htm

Hodges, C. (2004). Designing to motivate: Motivational techniques to incorporate in e-learning experiences. *The Journal of Interactive Online Learning, 2*(3), 1–7.

Johnson, R., & Onweugbuzie, A. (2004). Mixed methods research: A research paradigm whose time has come. *Educational Researcher, 33*(7), 14–26.

Jones, A., Issroff, K., Scanlon, E., Clough, G., & McAndrew, P. (2006). Using mobile devices for learning in informal settings: is it motivating? In *IADIS International Conference on Mobile Learning* (pp. 251–255). Dublin, Ireland: International Association for Development of the Information Society.

Keller, J. (1987a). Development and use of the ARCS model of instructional design. *Journal of Instructional Development, 10*(3), 2–10.

Keller, J. (1987b). Strategies for stimulating the motivation to learn. *Performance & Instruction, 26*(8), 1–7.

Keller, J. (1987c). The systematic process of motivational design. *Performance & Instruction, 26*(9–10), 1–8.

Keller, J. (2008a). An integrative theory of motivation, volition, and performance. *Technology, Instruction, Cognition, and Learning, 6*(2), 79–104.

Keller, J. (2008b). First principles of motivation to learn and e3-learning. *Distance Education, 29*(2), 175–185.

Keller, J., & Suzuki, K. (2004). Learner motivation and e-learning design: A mutinationally validated process. *Journal of Educational Media, 29*(3), 229–239.

Klassen, A., Creswell, J., Plano Clark, V., Smith, K., & Meissner, H. (2012). Best practices in mixed methods for quality of life research. *Quality of Life Research, 21*(3), 377–380.

Lawrence, E., Bachfischer, A., Dyson, L., & Litchfield, A. (2008). Mobile learning and student perspectives: An mreality check! Paper presented at the 7th International Conference on Mobile Business.

Lewin, K. (1935). *A dynamic theory of personality*. New York: McGraw-Hill.

Low, L., & O'Connell, M. (2006). Learner-centric design of digital mobile learning. *Online Learning and Teaching Conference – OLT Conference 2006* (pp. 71–82). Brisbane, Australia: Queensland University of Technology.

McMahon, M. (2010). *Evaluating the impact of digital media on apprenticeship training*. (University of Limerick, Department of Education and Professional Studies) Retrieved from http://ulir.ul.ie/bitstream/handle/10344/503/MM6mthesis.pdf?sequence=4

Murthy, S., Iyer, S., & Warriem, J. (2015). ET4ET: A large-scale faculty professional development program on effective integration of educational technology. *Educational Technology & Society, 18*(3), 16–28.

Naismith, L., Lonsdale, P., Vavoula, G., & Sharple, M. (2004). *Literature review in mobile technologies and learning.* Bristol, UK: NESTA Futurelab Series.

Pachler, N., Bachmair, B., & Cook, J. (2010). Charting the conceptual space. In N. Pachler, B. Bachmair, & J. Cook (Eds.), *Mobile learning structures, agency, practices* (pp. 3–27). New York: Springer.

Park, Y. (2011). A pedagogical framework for mobile learning: Categorizing educational applications of mobile technologies into four types. *International Review of Research in Open and Distance Learning, 12*(2), 78–102.

Patrichi, M. (2010). Mobile learning detractors. *Economy Informatics, 10*(1), 53–61.

Sagan, O., Waite, E., Cowan, S., Casey, H., & Karen, E. (2005). *Putting good practice into practice: literacy, numeracy and key skills in apprenticeships. Part two: Revisiting and reevaluating.* London: National Research and Development Centre for Adult Numeracy and Literacy.

Savage, M. (2003). Tackling the maths problem: Is it more extensive than we thought? Are solutions in sight? In S. Hibberd & L. Mustoe (Eds.), *IMA Conference Proceedings – Mathematical Education of Engineers IV* (pp. 127–131).

Sharples, M. (2000). The design of personal mobile technologies for lifelong learning. *Computers & Education, 34(3–4)*, 177–193.

Shellnut, B. (1996). *John Keller: Motivating influence in the field of instructional system design.* Retrieved from http://www.arcsmodel.com/pdf/Biographical%20Information.pdf

Sofaer, S. (2002). Qualitative research methods. *International Journal for Quality in Health Care, 14*(4), 329–336.

Stewart, D., Shamdasani, P., & Rook, D. (2006). *Focus groups: Theory and practice* (Vol. *20).* Thousand Oaks, CA: SAGE Publications.

Stockwell, G. (2008). Investigating learner preparedness for and usage patterns of mobile learning. *ReCALL, 20*(3), 253–270.

Stockwell, G. (2010). Using mobile phones for vocabulary activities: examining the effect of the platform. *Language Learning & Technology, 14*(2), 95–110.

Tolman, E. (1932). *Purposive behavior in animals and men.* New York: Century.

Tracy, F., & Carmichael, P. (2010). Research ethics and participatory research in an interdisciplinary technology-enhanced learning project. *International Journal of Research & Method in Education, 33*(3), 245–257.

Traxler, J. (2009). Learning in a mobile age. *International Journal of Mobile and Blended Learning, 1*(1), 1–12.

Traxler, J., & Kukulska-Hulme, A. (2015). Introduction to the next generation of mobile learning. In J. Traxler & A. Kukulska-Hulme (Eds.), *Mobile learning: The next generation.* London: Routledge.

Wang, Y., Wu, M., & Wang, H. (2008). Investigating the determinants and age and gender differences in the acceptance of mobile learning. *British Journal of Educational Technology, 40*(1), 92–118.

Whitehead, J., & McNiff, J. (2006). *Action research: Living theory.* London: Sage Publications.

Williams, B., & Bearman, M. (2008). Podcasting lectures: the next silver bullet. *Journal of Emergency Primary Health Care, 6*(3), 1–14.

Williams, P. (2009). Assessing mobile learning effectiveness and acceptance. (Unpublished Dissertation). The George Washington University, Washington D.C. ProQuest.

Wu, T., & Chao, H. (2008). Mobile e-learning for next generation communication environment. *International Journal of Distance Education Technologies*, *6*(4), 1–13.

Wyatt, T., Krauskopf, P., Gaylord, N., Ward, A., Huffstutler-Hawkins, S., & Goodwin, L. (2010). Cooperative m-learning with nurse practitioner students. *Nursing Education Perspectives*, *31*(2), 109–112.

Zhu, Q., Hu, Y., & Guo, W. (2012). Mobile learning in higher education: Students' acceptance of mobile learning in three top Chinese universities (Doctoral dissertation). Jönköping University, Sweden.

14

MOBILE LEARNING AND SOCIAL STUDIES HIGHER EDUCATION

A Thematic Review of Recent Research

Mark M. Diacopoulos

Introduction

In their position statement on the use of technology, the National Council for the Social Studies (NCSS) described a world where "mobile connectivity is interactive, instantaneous, and ubiquitous … global proliferation of cell phones offers testimony to a shift to mobile devices and the prospect of 24/7 connectivity for people worldwide" (NCSS, 2013). In recognizing the potential impact of mobile technology in social studies education, the NCSS challenged social studies educators of all levels to bring their "vision of powerful social studies education into the 21st century" (NCSS, 2013). Mobile learning (mlearning), as defined by Crompton (2013), is "learning across multiple contexts, through social contexts and interactions, using personal electronic devices" (p. 4). As an emerging education technology, mobile learning has the potential to form a vital part of the realization of the NCSS vision. However, little is known about how mobile learning technologies are being integrated and used in social studies education, and even less is known about the impact on learners when this technology is integrated in instruction, especially beyond the K–12 realm.

In order to better understand the use of mlearning in higher education, a thematic review was conducted. The main question used to drive this thematic summary of recent research was:

1. To what extent is mobile learning integrated in social studies higher education?

With the intent of revealing more meaningful insights, the following sub-questions were also addressed:

2. In studies involving mobile learning and social studies, what were the major research purposes, methodologies, and outcomes?
3. In studies involving mobile learning and social studies, what were the content areas, educational levels, and educational contexts?
4. In studies involving mobile learning and social studies, what were the mobile devices used and the geographical distribution of the studies?

Method

The literature search was modeled on systematic review models and meta-analysis of mobile learning (e.g. Crompton et al., 2016). It was conducted using both an electronic set of databases and manual searches of specific journals to ensure a more extensive scope. To safeguard a level of quality, only articles published in peer-reviewed journals were included in this review. By using peer-review publications, a level of research quality was guaranteed. Articles were retrieved from an electronic search of educational databases: EBSCOHOST, Elsevier Direct, ERIC, JSTOR, Proquest, Sage Journal On-line, and Wiley International Science. Manual searches were also conducted in journals that include a focus on social studies education, specifically: *Social Studies Research and Practice*, *Theory and Research in Social Education*, *Journal of Social Studies Research*, and *The Social Studies.* Additionally, a manual search was conducted on the social studies section of *Contemporary Issues in Social Studies Education.* Manual searches were also carried out on journals that include a focus on mobile learning, specifically the *International Journal of Emerging Technologies in Learning*, *E-Learning and Education* (*ELEED*), *Journal of Mobile Teaching*, *Institute of Electrical and Electronics Engineers* (*IEEE*), and the *International Journal of Mobile and Blended Learning.*

In both the electronic and manual searches, the following search terms were used: "social studies", "history", "geography", "civics", "government", "psychology", and "sociology" with "mobile learning", "m-learning", "handheld", "tablets", "iPad", "situated learning", and "digital learning". The former terms describe the various educational contents that make up social studies, while the latter terms were employed as they are often used to describe features of mobile learning.

Inclusion/Exclusion Criteria

Table 14.1 below shows the inclusion and exclusion criteria. For inclusion in this review, each study had to meet all of the inclusion and exclusion criteria. Figure 14.1 illustrates the process by which articles were excluded from the initial search.

The initial search retrieved 2,086 articles. The titles were reviewed for relevance and from these 2,022 were excluded. From the 64 remaining, 36 articles were further excluded as they did not address social studies education. One article was excluded as it was a conference proceeding, and a further 17 were excluded

TABLE 14.1 Inclusion and Exclusion Criteria

Inclusion Criteria	*Exclusion Criteria*
Learners use mobile devices	The mobile device must not be a laptop
Social studies is primary focus	The mobile device must not be netbooks
or History is primary focus	Education technology integration was primary focus (e.g. not focused on social studies content)
or Geography is primary focus	
or Civics/Government is primary focus	The article pertains only to K–12 education
or Sociology is primary focus	
or Psychology is primary focus	
Mobile devices are used for learning	
Article is peer reviewed and published in a journal	
Article is available to the public	

as they were about mobile learning in K–12 social studies and therefore outside of the parameters of this review. The remaining 11 articles were subjected to thematic analysis.

Data Analysis

The 11 peer-reviewed articles which merited inclusion in the review were subjected to an initial round of inductive analysis and open coding similar to that used in grounded theory in the first cycle (Corbin & Strauss, 1990). Table 14.2 shows examples of the initial coding process.

In the second cycle emerging themes were cross-checked with the initial codes as illustrated in Table 14.3. The emergent themes were then checked for consistency to produce the repeating themes that became more distinct throughout the process. Table 14.3 shows initial codes and repeating themes. Three distinct themes framed the findings of this review.

Findings

As a result of this thematic review three repeating themes were addressed in the literature. These findings help illustrate the extent that mobile learning is currently addressed in higher education social studies.

Mobile Learning in Higher Education Social Studies Supports Social Studies Skills

Jarvis, Dickie, and Brown (2013) and Jarvis, Tate, Dickie, and Brown (2016) both emphasized the importance of mobile learning in supporting social studies skills.

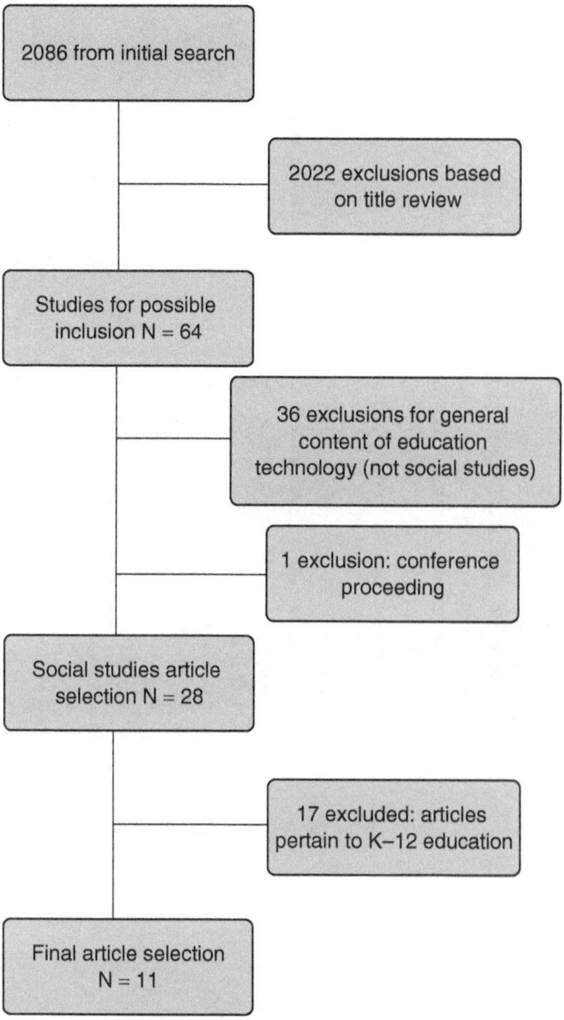

FIGURE 14.1 Flowchart of article selection

Both described evolving iterations of mobile learning used for geography fieldwork in a higher education context. They both emphasized the need for scaffolding in the context of fieldwork to enhance students' observation, inquiry, and thinking skills. Building on previous work involving the use of vodcasts in situ (Jarvis & Dickie, 2010), these studies investigated how mobile learning may be used to teach social studies skills. They concluded that as educators use mobile technologies to support context based teaching they must "reflect on what this

TABLE 14.2 Initial Codes

Article Number	*Author*	*Location*	*Content*	*Notes*	*Initial Codes*
1	Ormes (2015)	USA	Govt	Use of iPads in library. Helped with research. Mobile around library space.	Mobile device facilitates research in situ.
2	Chan et al. (2013)	Hong Kong	Psychology	Use of student response system on mobile devices. Use in classes increased engagement, learning, and understanding. Contingent on familiarity with software.	Mobile device for apps – Student Response System (SRS). Not mobile learning.
3	Stokes et al. (2012)	UK	Geography	Compared use of mobile devices in direct and remote access fieldwork. Direct access reported greater engagement and learning, as well as social emotional learning in group work. However, remote access to fieldwork produced similar learning outcomes. Authors argue that further potential for mobile devices in fieldwork can be leveraged.	Field data. In situ. Collaboration.
4	Kingston et al. (2012)	UK	Geography	Trialed use of mobile enhanced field experiences. Used GPS data, interactive video. Students enjoyed opportunity for interactivity and peer learning.	Field data. Multimedia. Collaboration.
5	Jarvis et al. (2016)	UK	Geography	Use of scaffolded activities around mediascapes to enhance in situ learning. Development of inquiry, observational, and thinking skills.	Multimedia. In situ. Inquiry.
6	Jarvis et al. (2013)	UK	Geography	Development of mediascape multimedia essay to assess in situ mobile fieldwork. Students created own learning.	Multimedia. In situ. Inquiry.
7	Jarvis and Dickie (2010)	UK	Geography	Describes using vodcasts in field-based settings to enhance field experiences.	Field data. Multimedia.

TABLE 14.3 Repeating Themes

Initial Codes	*Repeating Themes*
Field notes. Interactive content. Historical thinking.	Support social studies skills: Historical thinking, Thinking skills, Decision-making.
Augmented reality. Motivation. Engagement.	Student engagement.
Student centered learning. Field data. Collaboration.	Promotes collaboration.

significant shift in teaching method means for the way we access our students" (Jarvis et al., 2013, p. 89).

These articles similarly outlined how through a shift of pedagogical practice and an adoption of constructivist approaches, students can develop thinking skills, both geographic and historical, as well as engage in student-centered inquiry learning in situ. This affords educators the opportunity to utilize field-based instruction for more than data gathering on site, thereby modifying their existing practice. These provide examples of how the ability to interact, record, construct meaning, and create products using mobile devices can potentially transform the learning of social studies skills.

Mobile Learning in Social Studies Promotes Student Engagement

Ormes (2015) described how the use of iPads in a library space helped students to engage in research-based activities that went beyond their usual experiences in a library space. Similarly, Chan et al. (2013) described experiences of using mobile devices in social studies classrooms to improve engagement. Chan et al. contended that although students sometimes needed help to overcome issues of understanding the software, they were more willing to express opinions and participate in class activities when using mobile devices.

Extending the work of Ormes (2015) and Chan et al. (2013), in which mobile devices improved engagement in traditional learning spaces, Kingston et al.'s (2012) study explored the use of mobile learning devices for virtual field tours in a hydrology undergraduate course. They found that students were better engaged with the content through learning with these devices. Kingston et al. (2012) reported that using GPS-enabled devices taught students practical transferrable skills; however, the authors cautioned other researchers that although more engaging than learning in a traditional classroom environment, virtual field experiences should not replace the importance of making "first hand observations … because this is vitally important in understanding the complexity of environmental systems and how theoretical knowledge can be applied in a real world context" (p. 1286). This was consistent with Chan et al.'s contention that learning occurring in a field-based experience cannot be substituted with a mobile app. In this context, all three studies were similar, as they reported mobile learning experiences that intended to substitute, not transform, learning experiences.

Mobile Learning in Social Studies Supports Student Collaboration

Mobile learning also promoted opportunities for collaboration when used in geographic contexts, particularly when engaging in fieldwork. This was best demonstrated by Stokes et al. (2012) whose experimental design reported how, among other things, students felt that the level of social interaction afforded from the field experiences supported by mobile learning, was positive. The authors reported that student interactions with both tutors and each other was crucial to their learning experience. Interestingly, the experimental group of students who were given remote access to a field site reported greater collaboration as they felt a stronger need to communicate with each other to broaden their limited understanding of the site. For each of the groups participating in this particular study, uses of mobile learning, whether in situ or remotely, fostered collaborative learning.

Similarly, Chang et al. (2012) and Goh et al. (2012), investigated an app for collecting field data by trainee teachers. Both studies reported that participants expressed a need for more opportunity to communicate with each other in a more stable application than the one on offer in the study. The authors felt the app under investigation encouraged students to work together; however, it was not as sophisticated as other commercial, communication apps with which the students were more familiar. This finding has implications for those who design custom apps for social studies purposes. As students become familiar with the collaborative aspects of mobile devices through social media use, they expect a similar level of collaboration in their educative mobile learning apps. This isn't to say that mobile learning for fieldwork is not effective. A recent study by France et al. (2016) concluded that students who use and are aware of mobile apps for fieldwork display more attributes for future employment in their field. The implication of France et al.'s study is the use of mobile learning in the field not only supports student collaboration, but also has the potential to develop professional habits that may improve the likelihood of future employment.

Discussion

It is unclear from the review whether student collaboration is supported because of the situation in which learning is conducted, for example site visits in history or fieldwork in geography. For example, the studies that describe student collaboration were unable to attribute the phenomena to the technology or the pedagogical approach of the learning activity. For all three findings, it is clear that when mobile learning is implemented in social studies higher education contexts, it is often part of a shift toward a student-centered, constructivist pedagogical approach. However, mobile learning is not necessarily transformative in outcome. Therefore, factors such as student collaboration, engagement, and the overt teaching of social studies skills are important findings, as they are often already evident in current social studies higher education practice, but may be better supported through mobile learning. Most authors agreed on a need for more research into how constructivist pedagogy can be leveraged with mobile learning tools. Moreover, other studies point out a need for instructors to better understand what mobile learning is and how it can be used to enhance social studies instruction (Pocatilu et al., 2012). There is much more to investigate before the research community can declare the extent that mobile learning truly supports these aspects of social studies instruction. Further research is needed if we are to fully ascertain the potential of mobile learning to transform social studies teaching.

Moreover, as this systematic review demonstrated, even though the literature addressing mobile learning in social studies education is dominated by several repeating themes, the research base is severely limited. For the purposes of this review, there were only 11 articles that addressed mobile learning in social studies higher education contexts. So it is with caution that these emerging trends are reported. Several potential contexts were either under-reported or not reported at all. For example, there were very few studies addressing social studies content in higher education in areas other than geography-related contexts. If other content areas (history, sociology, psychology, government and political science, or religious studies) leverage mobile learning at the higher education level, then there is little published research to reflect it.

As this review demonstrated, the most significant findings were in the affective domain and addressed issues of student engagement and collaboration. This could be because the technology is still so relatively new. Consistent with other technology research in social studies, the focus is often on what the new technology can do, rather than what it really means to student attainment or outcomes. As a critique of social studies and technology research, this is consistent with earlier technological pedagogical trends in social studies (e.g. Swan & Hofer, 2008).

At this stage in its evolution, mobile learning in the social studies is relatively new, and therefore more research is needed in all aspects of its implementation. If social studies educators are to truly understand the complex nature of learning with this technology, and use it to improve pedagogical practice, then more needs to be done to discover whether learning with mobile devices really does improve student engagement beyond the novelty of a different tool for the same learning. Future researchers need to ascertain the reasons why students appear to collaborate with one another more successfully when learning in this way. How mobile learning can be applied to the broad range of the social sciences also merits investigation, as do the various contexts (beyond fieldwork) in which mobile learning is implemented. With the fast pace of technological change, the necessity to conduct research in this field is imperative in order to fully understand if mobile learning can realize its potential in social studies higher education.

Conclusions

This systematic review was conducted to ascertain the extent to which mobile learning is integrated in higher education social studies instruction. At the time of writing, it is accurate to infer that that mobile learning is in its infancy as a higher education social studies pedagogy. Examples of its use in geography-related contexts, usually as a way of facilitating field trips or fieldwork experiences, were reported. If mobile learning is present in the other areas of social sciences, at this time there is little to no research to reflect the extent of its use.

When investigating mobile learning in social studies, the majority of researchers adopted a case study approach, relying on qualitative data to report the experiences of the participants. Mixed methods and true experimental designs were sometimes used to measure the efficacy of a tool being tested, or to compare patterns of engagement between control and experimental groups. When results were reported, anecdotal reports often demonstrated the pedagogical approach used or the tools that were implemented in the study. Affective outcomes were also a focus of many of the studies (e.g. engagement, collaboration, student satisfaction). Few articles investigated student outcomes in their learning, and none reported anything significant pertaining to student attainment in their findings.

Of the 11 articles included in this review, eight were in geography-related contexts, one in government, one in psychology, and one in economics. Interestingly, there were no studies about mlearning in history at the higher education level. The geography articles generally focused on the conduct of fieldwork using mobile devices. This may be because the conduct of fieldwork is a key feature of geography education and is best suited to the mobile aspect of mobile learning.

Geographically, the 11 studies were from different parts of the world. The United States only accounted for two articles in this review, Hong Kong accounted for one, the United Kingdom contributed five articles mostly focused on geography fieldwork, Singapore two, and Romania one. A wide variety of devices were used, from mobile phones (both Android and iPhone were mentioned) to iPads. Of these the choice of device was made on the ability of users to access either commercial or custom-designed apps that could facilitate learning or data collection.

In addition, it is important to note the majority of articles reviewed demonstrated the ability of mobile learning to substitute or augment existing practice; for example: using mobile devices to collect and measure data as an augmentation of traditional fieldwork methods. Most studies speculated that in time mobile learning could be transformative for learners, unfortunately this was not a phenomena that was reported in the articles.

It is the expectation of the NCSS that technology will help achieve its vision for powerful social studies education (NCSS, 2016); furthermore, the NCSS recommends that pre K–16 educators should integrate technology into student learning because:

> the promotion of technology integration into student learning requires remaining sensitive not only to existing, but also to emerging, technologies so as to ensure that a learning environment is in place which best enables students to make rich use of those technologies as they prepare to become digital citizens.
>
> *NCSS, 2013*

As an emerging technology, instructors, educators, and researchers of social studies in higher education settings should be mindful of how mobile learning may help to create the best learning environment for our students. At this early stage in the research cycle, there is still much to be learned about mobile learning in social studies higher education. The studies included in this review have merely scratched the surface in trying to understand the potential of this approach to transform social studies education. It is necessary to call for more research in this field in order to gain a better understanding of not only how mobile learning is implemented at this level, but to gain a deeper understanding of the effects of mobile learning on student understanding in higher education social studies.

The current trends outlined in this review imply that mobile learning may be beneficial to improve student engagement and student collaboration if the learning is appropriately scaffolded around the apps or content under investigation. However, of concern to the author are the relatively small number of studies in this review. This begs the question: is the paucity of research on mobile learning in higher education social studies indicative of the limited opportunities for research in this field? Or, is mobile learning so new in social studies that there are simply not enough examples of mobile learning happening to report significant findings? The latter may be true in areas such as history in the higher education setting where there are articles that focus on the K–12 field but none in higher education. Alternatively, it could be asked whether the articles included in this review are merely the tip of a growing iceberg of research that is yet to surface. As this systematic review has demonstrated, there is yet more to learn about how mobile learning is implemented in higher education social studies.

References

Chan, K., Brown, I., Chung, I.C.B., Hui-Jing, L., & Luk, G.W.T., (2013) Using student response system via mobile devices in large introductory psychology class. In E. Ivala (Ed.), *Proceedings of the 8th International Conference of E-Learning* (pp. 76–81). Reading, UK: Academic Conferences and Publishing International Limited.

Chang, C.H., Chatterjea, K., Goh, D.H.L., Theng, Y.L., Lim, E.P., Sun, A., Razikin, K., Kim, T.N.Q., & Nguyen, Q.M. (2012). Lessons from learner experiences in a field-based inquiry in geography using mobile devices. *International Research in Geographical and Environmental Education, 21*(1), 41–59.

Corbin, J.M., & Strauss, A. (1990). Grounded theory research: procedures, canons, and evaluative criteria. *Qualitative Sociology*, *13*(1), 3–21.

Crompton, H. (2013). Mobile learning: New approach, new theory. In Z. L. Berge & L. Muilenburg (Eds.), *Handbook of mobile learning* (pp. 47–57). New York: Routledge.

Crompton, H., Burke, D., Gregory, K., & Gräbe, C. (2016). The use of mobile learning in science education: A systematic review. *Journal of Science Education and Technology, 25(2)*, 149–160.

France, D., Powell, V., Mauchline, A.L., & Park, J. (2016). Ability of students to recognize the relationship between using mobile apps for learning during fieldwork and the development of graduate attributes. *Journal of Geography in Higher Education*, *40*(2), 182–192.

Goh, D.H.L., Razikin, K., Lee, C.S., Lim, E.P., Chatterjea, K., & Chang, C.H. (2012). Evaluating the use of a mobile annotation system for geography education. *The Electronic Library*, *30*(5), 589–607.

Jarvis, C., & Dickie, J. (2010). Podcasts in support of experiential field learning. *Journal of Geography in Higher Education*, *34*(2), 173–186.

Jarvis, C., Dickie, J., & Brown, G. (2013). Going mobile: Perspectives on aligning learning and teaching in geography. *Journal of Geography in Higher Education*, *37*(1), 76–91.

Jarvis, C., Tate, N., Dickie, J., & Brown, G. (2016). Mobile learning in a human geography field course. *Journal of Geography*, *115*(2), 61–71.

Kingston, D.G., Eastwood, W.J., Jones, P.I., Johnson, R., Marshall, S., & Hannah, D.M. (2012). Experiences of using mobile technologies and virtual field tours in physical geography: Implications for hydrology education. *Hydrology and Earth Systems Sciences*, *16*, 1281–1286.

National Council for the Social Studies (NCSS). (2013). Technology position statement and guidelines. NCSS. Washington, DC. Retrieved from http://www.socialstudies.org/positions/technology

National Council for the Social Studies (NCSS). (2016). A vision statement for powerful social studies education. NCSS, Washington, DC. Retrieved from http://www.socialstudies.org/publications/socialeducation/may-june2016/vision-of-powerful-teaching-and-learning-in-social-studies

Ormes, D. (2015). Smart technology classrooms in the stacks: Promoting free government apps in an academic library. *Library Hi Tech News*, *1*, 13–17.

Pocatilu, P., Visoiu, A., Doinea, M, & van Osch, W. (2012). Mobile learning and mobile technologies in academia: A case study. *Economic Computation and Economic Cybernetics Studies and Research*, *46*(3), 79–98.

Stokes, A., Collins, T., Maskall, J., Lea, J., Lunt, P., & Davies, S. (2012). Enabling remote access to fieldwork: Gaining insight into the pedagogic effectiveness of 'direct' and 'remote' field activities. *Journal of Geography in Higher Education*, *36*(2), 197–222.

Swan, K.O., & Hofer, M. (2008). Technological pedagogical content knowledge in action: A case study of middle school digital documentary project. *Journal of Research on Technology in Education*, *41*(2), 179–200.

15

CULTURES OF MOBILE LEARNING

A Review of Learning Effectiveness in Mobile-Rich Classrooms

Michael Stevenson, Kumaran Rajaram and John Hedberg

In recent years, mobile technology has become widely available and increasingly sophisticated. Many in higher education (HE) advocate the use of smart and mobile devices to support and enhance learning. In spite of mobile learning (mlearning)'s potential, further research is needed to understand the intersections between the culture of learning within the institution and actual learning outcomes. Mlearning has been explored in a very wide range of cultural and institutional contexts, each of which presents unique challenges and opportunities. The learning benefits of mobile devices—such as greater self-efficacy, improved knowledge retention and superior forms of collaboration—frequently stand in contrast to disadvantages, such as increased distraction, greater inequity and the negative impact of software and hardware limitations on the learning experience.

Method

To understand and address how both effective and ineffective aspects of mlearning are situated in their broader institutional context, we performed a systematic review of empirical studies of mlearning in HE to date. Researchers formulated three research questions:

1. What are the attributes of effective and ineffective mlearning in HE?
2. What are the cultural factors that shape effective and ineffective mlearning in HE?
3. How, and in what ways, is mlearning shaped by these cultural factors?

Online thesaurus checks of linkages of "mlearning", "cultures of mobile-learning" and "higher education" were conducted. From here, Boolean searches for containers of key terms employed by several major databases to generate results were carried out. An initial set of 97 studies were assessed against selection criteria that included: (1) peer-reviewed studies of wireless and/or mobile digital devices in higher education; (2) studies with documented learning objectives related to the use of the technology in question; and (3) studies that provided sufficient cultural context about the use of the technology in question for the purpose of learning. Based on these criteria, 19 studies were omitted from the meta-analysis.

The resulting set of 68 studies comprised the meta-analysis for this review. One researcher coded each of the studies based on several attributes: *country/countries, study type, discipline area, sample size, instrumentation, theoretical framework, attributes of effective learning, attributes of ineffective learning, study context* and *further research needed*. Codes were independently checked by two other researchers for reliability and consistency. One researcher then synthesized empirical findings in relation to learning effectiveness by critically reviewing each study's reported findings, while a second researcher synthesized findings in relation to the cultural context of the institution by critically examining the context in which the study was carried out. Researchers then identified key themes drawn from the topics explored across the set of studies, the reported findings and the contexts in which the studies took place. Patterns such as agreement and disagreement across study findings were identified along with anomalies and limitations. Key demographic data pertaining to methodology and context were then enumerated to divulge further insights into the reported findings. Researchers then documented both the strengths and weaknesses in the available literature, noting key findings from the meta-analysis and possibilities for further research.

Results

Methodology, Instrumentation and Sample

The set included studies that were purely quantitative (n = 32, 47.1 per cent), purely qualitative (n = 12, 17.6 per cent), and those that employed mixed methodologies (n = 22, 32.4 per cent), with two meta-analytical studies also included in the analysis (2.9 per cent) but excluded from the sample. The instrumentation employed across the set included studies that utilized single- (n = 49, 73.1 per cent) and multi-method approaches (n = 18, 26.9 per cent). The single method group included a majority of studies solely employing a questionnaire (n = 26), with remaining studies employing content analysis (n = 13), one-on-one interviews (n = 6), focus group interviews (n = 2) and meta-analysis (n = 2). The multi-method group collectively employed 38 instruments, comprising questionnaires (n = 14), content analysis (n = 11), one-on-one interviews (n = 11) and focus group interviews (n = 2).

Figure 15.1 shows the breakdown of instruments used across the reviewed literature. Questionnaires were commonly employed in both quantitative and mixed

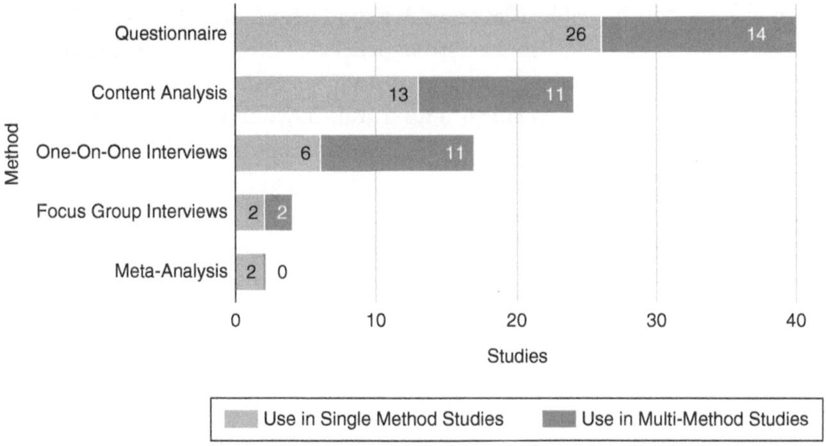

FIGURE 15.1 Use of Instrumentation in Studies Reviewed.

studies, while the 24 studies employing various forms of content analysis were spread across quantitative (n = 7), qualitative (n = 4) and mixed methodologies (n = 13).

The total sample size of all studies reviewed was 10,487, which included larger combined samples in the quantitative (n = 8257, M = 258, SD = 304) and mixed studies (n = 1820, M = 83, SD = 67), with a smaller combined sample across the qualitative studies (n = 410, M = 34, SD = 41). Of note, the very high standard deviation for the combined quantitative sample reflected the wide variance in the samples reported across all studies in this category. Of the studies reviewed, a majority focused on undergraduate students (n = 43, 63.2 per cent), with smaller portions focusing on graduate students (n = 10, 14.7 per cent) and a mix of both undergraduate and postgraduate students (n = 14, 20.6 per cent).

Discipline, Research Topic and Theoretical Framework

Exactly half the studies reviewed (n = 34, 50 per cent) did not specify a discipline area in which the study has being conducted. There were smaller groups in Languages (n = 6, 8.8 per cent), Science and Business (both n = 4, 5.9 per cent), Psychology and Information Technology (both n = 3, 4.4 per cent) and Research and Geography (both n = 2, 2.9 per cent). Remaining studies were unique in the discipline area they addressed, with disciplines that included Mathematics, Management, Instructional Design, Ecology, Journalism, Statistics, Sociology, Art and Design, Nursing and Education. These disciplines collectively represented only 13.2 per cent of the studies reviewed.

20 topics were derived from the set of reviewed studies that reflect the intersection of mlearning with other areas of elearning. The topics explored were, by and large, unrelated to study disciplines. The exceptions to this finding were studies exploring mobile devices for writing and publishing in Journalism classes

(Kraft & Seely, 2015), graphic design in Art and Design (Souleles, Savva, Watters, Annesley, & Bull, 2015), 3D representations in Science (McCollum, Regier, Leong, Simpson, & Sterner, 2014) and augmented reality in Nursing classes. Other topics fell into non-discipline specific groups that included instructional uses such as incorporating podcasting (n = 6, 9.1 per cent), flashcards (n = 1, 1.5 per cent) and mobile blogging (n = 3, 4.5 per cent) to strengthen and deepen understanding of course content. There was a small group of studies (n = 5, 7.6 per cent) exploring more holistic pedagogical approaches that included project-based learning, cooperative learning, seamless learning, games-based learning and fieldwork. The final group (n = 5, 7.6 per cent) consisted of studies that focused on IT infrastructure to facilitate mlearning in areas such as Learning Management Systems, distance education and instructional design.

Table 15.1 shows the list of research topics derived from the literature review, with frequencies for each topic. In some cases, the topic was clearly conveyed in the study's methodology and findings, and the codification of topics was relatively clear. However, 28 studies chose to focus on more generalized uses of mobile devices for learning, without specifying software, websites or other applications. To reflect the focus of these studies, the code "personalized learning" was employed.

In relation to theoretical frameworks, researchers observed that a large subset of the studies reviewed did not report a theoretical framework (n = 33, 48.5 per cent).

TABLE 15.1 Mlearning Topics Explored across Reviewed Literature

Topic	*Number of Studies*	*Coverage (%)*
Personalized Learning	28	42.4%
Instant Messaging	10	15.2%
Podcasting	6	9.1%
Mobile Blogging	3	4.5%
Learning Management Systems	3	4.5%
Research	2	3.0%
Social Media	1	1.5%
Flashcards	1	1.5%
Problem-Based Learning	1	1.5%
Instruction	1	1.5%
Augmented Reality	1	1.5%
Distance Education	1	1.5%
Group Work	1	1.5%
Games-Based Learning	1	1.5%
Writing and Publishing	1	1.5%
Seamless Learning	1	1.5%
Representational Competence	1	1.5%
Graphic Design	1	1.5%
Process Management	1	1.5%
Fieldwork	1	1.5%

TABLE 15.2 Breakdown of Theoretical Frameworks in Reviewed Literature

Theory	*Number of Studies*	*Coverage (%)*
None	33	48.5%
Socio-constructivism	5	7.4%
Unified Theory of Acceptance	5	7.4%
Technology Adoption Model (TAM)	5	7.4%
Mlearning	4	5.9%
Problem-Based Learning	3	4.4%
Connectivism	2	2.9%
Activity Theory	2	2.9%
Cooperative Learning	1	1.5%
Affordance Analysis	1	1.5%
Reasoned Action and Planned Behavior	1	1.5%
Adult Learning Theory	1	1.5%
Distributed Cognition	1	1.5%
Self-Regulated Framework	1	1.5%
Digital Fluency	1	1.5%

Remaining studies adopted a range of frameworks, with most convergence in the *Unified Theory of Acceptance*, *Technology Adoption Model* and *Socio-Constructivism* (all n = 5, 7.4 per cent). Table 15.2 shows the breakdown of theoretical frameworks explored in the literature.

Improved Learning Outcomes

Across the set of studies reviewed, researchers noted nine key areas where there was evidence of improved learning outcomes. Of these, a large portion of studies reported *positive perceptions* of mobile devices (n = 27, 40.9 per cent) as the main outcome of the research.

Findings in one group of studies suggest that prior experience with mlearning is essential in accurately identifying its advantages over other forms of learning. In their investigation of mlearning in distance education, Han and Han (2014) found that both users and non-users could identify advantages and disadvantages. Alyahya and Gall (2015) similarly found that students were generally aware of the learning affordance of mlearning and able to employ these affordances in their own learning. Gikas and Grant (2013) examined perceptions of social media in mlearning, finding that students were able to identify both advantages of using the tools and frustrations with app and device limitations. However, in a study of students' use of iPads for fieldwork, Welsh (2015) observed that although some students viewed the device as a "multi-tool" for learning in the field, they did not accurately identify its other uses. In a comparative study of smartphones, tablets and laptops for learning, Anand (2014) reported a sizeable portion of students preferring traditional laptops over newer mobile devices. Three studies also independently found gender to be a strong predictor of use, with significantly more male users reporting positive perceptions than their female counterparts (Baker, Lusk, & Neuhauser, 2012; Beaver, Knox, & Zusman, 2010; Tajudeen, Basha, Michael, & Mukthar, 2013). Summarizing the findings, Tajudeen (2013) asserts that the "gender issue is crucial for the adoption of mobile devices in learning" (p. 26).

A number of studies (n = 12, 18.2 per cent) reported *improved skills* as a key outcome. The improved skills reported across these studies included superior language development (Basoglu & Akdemir, 2010; Huang, 2014), improved learner inquiry and problem-solving (Utulu & Alonge, 2012), superior note-taking (Khechine, Lakhal, & Pascot, 2013), more effective revision (McConatha, Praul, & Lynch, 2008) and improved content recall (Kuznekoff, Munz, & Titsworth, 2015). Studies by Luna and Cullen (2011) and McKinney (2009) found that podcast analysis could substantially deepen knowledge of course content. However, only two studies (3 per cent) reported *higher order thinking* as an improved learning outcome. Interestingly, both studies employed theoretical frameworks that support higher order thinking. Adams and Blair (2014) found that Cooperative Learning could be employed in multimodal analysis of podcasts for superior higher order thinking, while Chan et. al. (2015) showed that Problem-Based Learning (PBL) was essential for using mobile devices to support critical thinking.

For the three studies (4.5 per cent) reporting *improved communication*, there is evidence to suggest that instant messaging can facilitate better communication between teachers and learners. Hsu and Ching (2012), Liu, Chung, Chen and Liu (2009) and Macgregor, Spiers and Taylor (2011) all found that mobile communication benefits feedback, questioning and note-taking. However, Liu, Chung, Chen and Liu (2009) caution that the small screens of most mobile devices, although effective for all forms of communication, do not effectively support group work.

Both Arpaci (2015) and Ezemenaka (2013) report *increased adoption* as a result of mlearning being implemented in higher education. Arpaci (2015) studied the adoption of mlearning in both Canada and Turkey, finding that there were significant cultural differences in play, with social influence the strongest predictor of effective use in Turkey and innovativeness the strongest predictor in Canada. Studying the same contexts, Tekinarslan, Kennedy and Nicolle (2015) found that US students showed much greater uptake of mlearning, with better access to devices and greater acceptance of their use in education. However, in a similar comparative study of Nigeria and the USA, Itegboje (2015) found that reliance on mobile devices was significantly higher in Nigeria—as a developing country—but that disparity of wealth meant that US students reported higher levels of perceived usefulness with devices that were more capable than their Nigerian counterparts.

Several studies (n = 8, 12.1 per cent) reported *increased engagement* in mlearning as a key learning outcome. In terms of the individual learner's engagement, Abdullah, Siraj and Hussin (2014) reported more engaged language learners, while

Kinash, Brand and Mathew (2012) found that the device itself—the iPad—could "motivate" learners. Studies by Cobb, Heaney, Corcoran and Henderson-Begg (2010), Geng (2013) and Tremblay (2010) all found that mobile-based instant messaging led to higher levels of learner engagement, with Cobb et al. (2010) further noting that these forms of learning appear to better support shy and introverted learners. Cochrane and Bateman (2011) found that where a community of practice in relation to mlearning is developed, more consistent levels of engagement are fostered.

Finally, three studies (4.5 per cent) reported *improved self-regulated learning*. Garrett, Jackson and Wilson (2015) studied mlearning with augmented reality, finding that both the learner's reported levels of comfort and ease of use strongly correlated with higher levels of self-directed learning. Sung et al. (2014) piloted a prompt-based approach to mlearning, finding that where appropriate prompts are issued, there are higher levels of reported self-efficacy. Voelkel and Bennett (2014) found that the use of instant message polling during in-class facilitations and presentations led to higher levels of metacognition, with learners more aware of their weaknesses than with a lecture-only approach.

Ineffective Learning Outcomes

Significantly, seven studies (10.6 per cent) reported limited or no improvement to learning. In the findings of studies by Burns and Lohenry (2010), La Roche and Flanigan (2013) and Elder (2013), learner distraction appeared to outweigh the perceived benefits. Elaborating on their findings, La Roche and Flanigan (2013) draw attention to distraction as an impediment to higher order thinking, such as problem-solving and critical thinking:

> While note-taking on a laptop is arguably a positive form of engagement, Facebook, E-mail, and Twitter are at the very least a form of retreatism and when they create a distraction to the professor and those around them, these activities are a form of rebellion. Looking up a resource could be a positive form of engagement or possibly a replacement for critical thinking.
>
> *(p. 49)*

Studies by Duygu-Eristi, Haseski, Uluuysal and Karakoyun (2011) and Wang, Wiesemes and Gibbons (2012) found that hardware limitations of mobile devices can negatively impact on learning experiences and outcomes. Both studies cite small screens, connectivity problems and file access as factors. In their study of learner engagement with mlearning, La Roche and Flanigan (2013) found that the devices were far more likely to lead to distraction than engagement. While Souleles, Watters and Annesley (2015) observed that when given iPads, Art and Design students did not adequately explore the available apps for creativity, preferring paper-based learning instead.

Principally, the findings presented by Ezemenaka (2013), Kinash, Brand and Mathew (2012) and Nguyen, Barton and Nguyen (2015) suggest that gains in learner engagement and positive perceptions of the devices and/or the learning experience do not consistently translate into improved learning outcomes. Ezemenaka (2013) concludes that "there is no significant relationship affecting the academic performance of students in using internet enabled phones" (p. 171). In total, six studies (9.1 per cent) noted the impact of distraction on the learner, with Gasaymeh and Qablan (2013) also referring to "mobile addiction" as a serious impediment. 14 studies (20.6 per cent) cited the hardware limitations of mobile devices as a factor predicting ineffective learning. Five of these studies (6.1 per cent) specifically cited screen size as the main concern, while in two of the contexts studied (3 per cent), poor wireless connectivity was the main factor that predicted poor learning outcomes.

Discussion

Jubien (2013) summarizes perhaps the biggest challenge faced by those in higher education seeking to implement and support effective mlearning: "If universities do not own smartphones, then they have less influence and control about how they will be used in education" (p. 10). The highly personal nature of mobile devices ultimately means that there can be few, if any, one-size-fits-all approaches.

The analysis of qualitative data occurred in thirty-four of the studies reviewed. Further qualitative and mixed method studies are needed to capture the breadth and depth of the learner's experience of mlearning while conveying the cultural context. The lack of a specific discipline area in half of the studies reviewed represents a limiting factor for identifying improved learning outcomes, since studies that explore general uses unrelated to discipline content can only identify generalized outcomes that require further application to discipline areas. Further studies are also needed that illuminate findings about pedagogically informed implementation of mlearning in key discipline areas, such as Marín et al. (2016), who explored the use of mobile devices in seamless learning environments through the application of Project-Based Learning. Similarly, the absence of pedagogy as a consideration in many of the studies reviewed means that conclusions about improved learning outcomes are tenuous, where these conclusions do not clearly explain the role of pedagogy.

Our review suggests that mlearning intersects with a range of other technologies, some of which exist independently from mobile device use. Further research is needed to explore interaction effects between these technologies and the use of mobile devices. Social media, the LMS, Web 2.0, SMS and augmented reality have their own bodies of literature with important findings. Further research might consider how these technologies have further evolved with mlearning and draw comparisons between app-based learning on mobile platforms and traditional computing on laptop and desktop computers. Such research may address the problem of *generativity* posed by critics of mobile device use such as Zittrain (2009).

The very large number of studies specifying no theoretical framework (n = 33, 48.5 per cent) meant that these studies' findings may not effectively build our understanding of mlearning. Some of the frameworks discussed reflect much older theories that predate the technologies being explored, while newer developmental theories such as *Connectivism* (Siemens, 2014) and the *Unified Theory of Acceptance Model* (Venkatesh et al., 2003) show potential for the development of a learning theory that adequately explains effective mlearning. Regardless, the inclusion of a theoretical framework in all of the studies strongly correlated with the identification of actual improved learning outcomes.

Finally, our review suggests that there remains a preoccupation with exploring learners' perceptions of mobile devices for learning. *Learner perceptions* have been—and remain—a very important factor to consider when exploring the implementation of a new technology within any learning context. Furthermore, these studies suggest that the perceptions themselves can act as predictors of engagement, adoption and effective use. However, mobile devices are now ubiquitous, and there is arguably a pressing need to move beyond our investigation of the learners' perceptions towards a better understanding of their actual uses of the devices for learning and whether these uses make a meaningful difference to actual learning outcomes.

Conclusion

The findings of this review suggest that it is possible to draw from the literature evidence of effective and ineffective mlearning. There is some evidence to support purported benefits of mlearning, such as higher levels of collaboration, enhanced self-efficacy and improved knowledge retention. In contrast, though, there are reported disadvantages such as widening gaps in participation, gender and equity, higher levels of distraction and the negative impact of hardware and software limitations on the learner's experience. These aspects represent important cultural factors contingent on institutional context and on how learners perceive and interpret the learning experience. While the general characteristics of effective mlearning can be collated and replicated, findings need to be critically and objectively verified in terms of the levels of effectiveness when they are adopted through varying cultural contexts. This would allow further calibration of the identified effectiveness factors to be more tailor-made to the specific context based on the learners' preferences, learning approach, and inclination of how the knowledge is to be transferred. Ineffective mlearning could, likewise, be explored more closely in relation to the cultural context, and should not necessarily be generalized.

References

Abdullah, M. R. T. L., Siraj, S., & Hussin, Z. (2014). Interpretive structural modeling of mlearning curriculum implementation model of English language communication skills for undergraduates. *Turkish Online Journal of Educational Technology, 13*(1), 151–161.

Adams, R. V., & Blair, E. (2014). The learner-generated podcast: Engaging postgraduate engineering students in a mathematics-intensive course. *Research in Post-Compulsory Education, 19*(2), 132–46.

Anand, G., Chhajed, D., Hong, S.W., & Scagnoli, N. (2014). Introducing mobile technology in graduate professional education. *E-Learning and Digital Media, 11*(6), 543–553.

Arpaci, I. (2015). A comparative study of the effects of cultural differences on the adoption of mobile learning. *British Journal of Educational Technology, 46*(4), 699–712.

Baker, W. M., Lusk, E. J., & Neuhauser, K. L. (2012). On the use of cell phones and other electronic devices in the classroom: Evidence from a survey of faculty and students. *Journal of Education for Business, 87*(5), 275–289.

Basoglu, E. B., & Akdemir, O. (2010). A comparison of undergraduate students' English vocabulary learning: Using mobile phones and flash cards. *Turkish Online Journal of Educational Technology, 9*(3), 1–7.

Beaver, T., Knox, D., & Zusman, M. E. (2010). "Hold the phone!": Cell phone use and partner reaction among university students. *College Student Journal, 44*(3), 629–632.

Begum, R. (2011). Prospect for cell phones as instructional tools in the EFL classroom: A case study of Jahangirnagar University, Bangladesh. *English Language Teaching, 4*(1), 105–115.

Burns, S. M., & Lohenry, K. (2010). Cellular phone use in class: Implications for teaching and learning a pilot study. *College Student Journal, 44*(3), 805–810.

Chan, L. K., Bridges, S.M., Doherty, I., Ng, L.N., Jin, J., Sharma, N., Chan, N.K., & Yan Yu Lai, H. (2015). A qualitative study on how health professional students and their PBL facilitators perceive the use of mobile devices during PBL. *Interdisciplinary Journal of Problem-Based Learning, 9*(1). Available from https://doi.org/10.7771/1541-5015.1510

Cobb, S., Heaney, R., Corcoran, O., & Henderson-Begg, S. (2010). Using mobile phones to increase classroom interaction. *Journal of Educational Multimedia and Hypermedia, 19*(2), 147–157.

Cochrane, T., & Bateman, R. (2011). Strategies for mlearning integration: Evaluating a case study of staging and scaffolding mlearning integration across a three-year bachelor's degree. *Waikato Journal of Education, 16*(1), 107–122.

Duygu Eristi, S., Haseski, H. I., Uluuysal, B., & Karakoyun, F. (2011). The use of mobile technologies in multimedia-supported learning environments. *Turkish Online Journal of Distance Education, 12*(3), 130–141.

Elder, A. D. (2013). College students' cell phone use, beliefs, and effects on their learning. *College Student Journal, 47*(4), 585–592.

Ezemenaka, E. (2013). The usage and impact of internet enabled phones on academic concentration among students of tertiary institutions: A study at the University of Ibadan, Nigeria. *International Journal of Education & Development Using Information & Communication Technology, 9*(3), 162–173.

Gall, J. (2015). Digital scribes: A qualitative study of graduate students who brought iPads into their academic lives. *Journal of Educational Multimedia and Hypermedia, 24*(4), 385–401.

Garrett, B. M., Jackson, C., & Wilson, B. (2015). Augmented reality m-learning to enhance nursing skills acquisition in the clinical skills laboratory. *Interactive Technology and Smart Education, 12*(4), 298–314.

Gasaymeh, A.M., & Qablan B.M. (2013). SMS as out-of-class, student-instructor interaction tool: A case study of Jordanian graduate students' perceptions and usage. *International Education Studies, 6*(8). Available from http://dx.doi.org/10.5539/ies.v6n8p147

Geng, G., & Disney, L. (2014). Exploring pre-service teachers' knowledge of and ability to use text messaging. *Australian Journal of Teacher Education, 39*(7). Available from http://dx.doi.org/10.14221/ajte.2014v39n7.4

Gikas, J., & Grant, M.M. (2013). Mobile computing devices in higher education: Student perspectives on learning with cellphones, smartphones & social media. *The Internet and Higher Education, 19*, 18–26.

Han, I., & Han, S. (2014). Adoption of the mobile campus in a cyber university. *International Review of Research in Open and Distance Learning, 15*(6), 237–256.

Hsu, Y.-C., & Ching, Y.-H. (2012). Mobile Microblogging: Using Twitter and mobile devices in an online course to promote learning in authentic contexts. *International Review of Research in Open and Distance Learning, 13*(4), 211–227.

Huang, R.-T. (2014). Exploring the moderating role of self-management of learning in mobile English learning. *Educational Technology & Society, 17*(4), 255–267.

Itegboje, A. O. (2015). Mobile technology adoption for teaching activities: Influencing factors in selected universities in Nigeria and the United States. Southern U and Agricultural and Mechanical Coll., US.

Jubien, P. (2013). Shape shifting smart phones: Riding the waves in education. *Canadian Journal of Learning and Technology, 39*(2), 16.

Khechine, H., Lakhal, S., & Pascot, D. (2013). University students' perception of the pedagogical use of podcasts: A case study of an online information system course. *Journal of Education and Training Studies, 1*(2), 136–151.

Kinash, S., Brand, J., & Mathew, T. (2012). Challenging mobile learning discourse through research: Student perceptions of "Blackboard Mobile Learn" and "iPads." *Australasian Journal of Educational Technology, 28*(4), 639–655.

Kraft, N., & Seely, N. (2015). Making mojos: How iPads are enhancing mobile journalism education. *Journalism & Mass Communication Educator, 70*(3), 220–234.

Kuznekoff, J. H., Munz, S., & Titsworth, S. (2015). Mobile phones in the classroom: Examining the effects of texting, Twitter, and message content on student learning. *Communication Education, 64*(3), 344–365.

La Roche, C. R., & Flanigan, M. A. (2013). Student use of technology in class: Engaged or unplugged? *Journal of College Teaching & Learning, 10*(1), 47–54.

Liu, C.-C., Chung, C.-W., Chen, N.-S., & Liu, B.-J. (2009). Analysis of peer interaction in learning activities with personal handhelds and shared displays. *Educational Technology & Society, 12*(3), 127–142.

Luna, G., & Cullen, D. (2011). Podcasting as complement to graduate teaching: Does it accommodate adult learning theories? *International Journal of Teaching and Learning in Higher Education, 23*(1), 40–47.

McCollum, B. M., Regier, L., Leong, J., Simpson, S., & Sterner, S. (2014). The effects of using touch-screen devices on students' molecular visualization and representational competence skills. *Journal of Chemical Education, 91*(11), 1810–1817.

McConatha, D., Praul, M., & Lynch, M. J. (2008). Mobile learning in higher education: An empirical assessment of a new educational tool. *Turkish Online Journal of Educational Technology*, 7(3), 15–21.

Macgregor, G., Spiers, A., & Taylor, C. (2011). Exploratory evaluation of audio email technology in formative assessment feedback. *Research in Learning Technology, 19*(2), 39–59. doi:10.1080/09687769.2010.547930

McKinney, D., Dyck, J. L., & Luber, E. S. (2009). iTunes University and the classroom: Can podcasts replace professors? *Computers & Education, 52*(3), 617–623.

Marín, V. I., Jääskelä, P., Häkkinen, P., Juntunen, M., Rasku-Puttonen, H., & Vesisenaho, M. (2016). Seamless learning environments in higher education with mobile devices and examples. Retrieved from https://jyx.jyu.fi/dspace/handle/123456789/48610.

Nguyen, L., Barton, S. M., & Nguyen, L. T. (2015). iPads in higher education-Hype and hope. *Journal of Educational Technology, 46*(1), 190–203.

Siemens, G. (2014). Connectivism: A learning theory for the Digital Age. Retrieved from http://er.dut.ac.za/handle/123456789/69

Souleles, N., Savva, S., Watters, H., Annesley, A., & Bull, B. (2015). A phenomenographic investigation on the use of iPads among undergraduate art and design students. *British Journal of Educational Technology, 46*(1), 131–141.

Sung, H.-Y., Hwang, G.-J., Liu, S.-Y., & Chiu, I. (2014). A prompt-based annotation approach to conducting mobile learning activities for architecture design courses. *Computers & Education, 76*, 80–90.

Tajudeen, S. A., Basha, M. K., Michael, F. O., & Mukthar, A. L. (2013). Determinant of mobile devices acceptance for learning among students in developing country. *Malaysian Online Journal of Educational Technology, 1*(3), 17–29.

Tekinarslan, E., Kennedy, E., & Nicolle, P. S. (2015). A cross-cultural comparative study of uses and perceptions of technology in education among Turkish and US undergraduates. *Journal of Education and Training Studies, 3*(6), 71–82.

Tremblay, E. A. (2010). Educating the mobile generation — using personal cell phones as audience response systems in post-secondary science teaching. *Journal of Computers in Mathematics & Science Teaching, 29*(2), 217–227.

Utulu, S. C., & Ayodele, A. (2012). Use of mobile phones for project based learning by undergraduate students of Nigerian private universities. *International Journal of Education & Development Using Information & Communication Technology, 8*(1), 4–15.

Venkatesh, V., Morris, M.G., Davis, G.B., & Davis, F.D. (2003). User acceptance of information technology: Toward a unified view, *MIS Quarterly, 27*(3), 425–478.

Viberg, O., & Gronlund, A. (2013). Cross-cultural analysis of users' attitudes toward the use of mobile devices in second and foreign language learning in higher education: A case from Sweden and China. *Computers & Education, 69*, 169–180.

Voelkel, S., & Bennett, D. (2014). New uses for a familiar technology: Introducing mobile phone polling in large classes. *Innovations in Education and Teaching International, 51*(1), 46–58.

Wang, R., Wiesemes, R., & Gibbons, C. (2012). Developing digital fluency through ubiquitous mobile devices: Findings from a small-scale study. *Computers & Education, 58*(1), 570–578.

Welsh, K. E., Mauchline, A. L., Powell, V., France, D., Park, J. R., & Whalley, W. B. (2015). Student perceptions of iPads as mobile learning devices for fieldwork. *Journal of Geography in Higher Education, 39*(3), 450–469.

Yarmey, K. (2011). Student information literacy in the mobile environment. *EDUCAUSE Quarterly, 34*(1).

Zittrain, J. (2009). *The future of the internet – and how to stop it*. New Haven, CT: Yale University Press.

CONCLUSION

Mobile Learning in Higher Education: The Challenges in Context

John Traxler and Helen Crompton

Introduction

This book looks beyond the clear and obvious achievements of the global mobile learning (mlearning) research community, achievements in exploiting progressively more powerful handsets and infrastructure to extend, enhance and enrich learning in higher education. It attempts to analyze and answer some of the more difficult questions, and it does so by looking at the experiences of our contributors in the contexts of global higher education.

Looking back, we can see that mobile learning was the focus for many innovative and thought-provoking projects and pilots, most of which took place in the institutions of higher education, and often involved students and teachers in higher education. While these pilots and projects were often spectacularly successful in terms of pedagogy and technology, they were less so in terms of scale, sustainability, evidence, impact, equity and embedding within the wider institutional settings and in terms of becoming established within the professional mainstream, the institutional framework and the quality and curriculum regulations. These are not new observations, and researchers have attempted to provide an improved understanding (Elgort, 2005; Ellis & Goodyear, 2013; Lee et al., 2011; Ng & Cumming, 2015; Romiszowski, 2004; Salmon, 2005) but usually only inside the bounds of the institutions (Herrington & Herrington, 2007; Herrington et al., 2009).

Meanwhile, mobile technology was becoming a universal, popular and social phenomenon, allowing people to generate, share, distribute, transform, discuss and own their own opinions, images, ideas and information—in fact, to become each other's teachers and learners, challenging the institutions of higher education to stay relevant and useful, to keep up with social and economic trends. Ideas of Bring Your Own Device (BYOD), user-generated content, and the flipped classroom were sometimes espoused but never fully enacted, and the legitimacy and authority of higher education's institutions started to look problematic. These issues are vitally important to the future of higher education and the current volume contributes to an informed and critical debate.

These are becoming challenging times for higher education and the mobile learning community must negotiate a credible role in the complex and fluid ecology of the universities and colleges of higher education in the many different parts of the world. This book maps this territory and includes a host of rigorous, scholarly, critical, and perhaps controversial contributions. It is customary in concluding chapters to look for themes and topics emerging out of such contributions. Apart, however, from arguing that these contributions are an artefact of editorial choices that grew out of criteria that included the quality of the writing and the spread of settings, to do so would be to fall into one of the abiding errors of innovative elearning over the last two decades. This is the error of moving from a handful of disparate instances to some upbeat generalization, to joining up too few dots and thinking we have learnt something new. In fact, this error is probably compounded of several component errors: the first is a consequence of the various forces that reinforce the reporting of success and resist the reporting of failure, the second is the selection of factors spuriously identified as causes and effects in this reporting, and the third is ignoring the separation in time, place and culture between the dots that we so wilfully try to join. These may, however, all be a consequence of the preference for evidence over expertise and experience, an aspect of the scientism of the twentieth-century zeitgeist. So, the challenge to the reader in reading these contributions is to make their own judgments, accepting that there are no easy answers and not always obvious connections. Perhaps in the case of learning with mobiles, history should teach how little history should teach us!

So, how then can we summarize where we are? What analytic tools would help?

The Contexts and the Challenges

We can perhaps talk about where we are in terms of concentric circles; the innermost contains the details of mobile learning innovation, pedagogy and technol ogy, outside of which are the various strands of innovation, trends and change in educational technology, themselves contained within the professions, the curricula, the missions, and the practices of the universities and colleges that constitute higher education, national, international and global. These do, however, exist within a wider, complex and varied technological, social, economic, political and cultural environment. This should, however, be a dynamic, fluid and permeable depiction where change in one circle, or annulus, affects those circles, or annuli, further out and further in.

The Outer Circle—The Outside World

Starting at the outermost circle, we have identified a couple of constituents that make up the environment of the global higher education system. Technology, the first one, is seemingly easier to discuss. Any simplistic analysis tells us that digital technology gets smoothly and inexorably better, small, faster and cheaper (Kumar, 2004). In the broad sense, this does seem to be true but any smooth curves and established paradigms have inevitably been disturbed by each new *game-changing* innovation, the iPhone being the most recent and obvious (Furfie, 2010; Wortham, 2009). Both general trends and game-changers are, however, driven by the commercial imperatives of the makers and the vendors of mobile digital technology. This historical trajectory, which puts more and more technology in the hands of individual consumers, is set to continue because it is a consequence of global competitive market forces, and so too, as a consequence, is the erosion of the institutional monopoly of digital technology. Looking backwards, there was a sea change when the balance of individual digital technology and institutional digital technology tipped away from the institutions, but the consequences of this are still to be felt. We currently live in a world of zombie universities, dead but still moving, unaware that individuals with persona and social technologies have taken their role, their rationale and their place.

The other major constituent of the outer circle is the general global economic and political environment and its capacity to fund the continued growth of education, of higher education and of digital technology within higher education. Like technology itself, but perhaps less confidently, we should be able to predict continued growth. This is, however, threatened by periodic crises, such as those precipitated by the sub-prime mortgage failures of 2008 (Demyanyk & Van Hemert, 2011); currently also by a marked movement towards neo-liberalism across much of North America and Western Europe (Hill & Kumar, 2012; Lakes & Carter, 2011; Olssen & Peters, 2005); and by the gradual drift of power, resources and capital away from North America and Western Europe and towards variously the economies of the Gulf States, the global South and the Pacific Rim, and the emerging economies of Russia, Brazil, India, China and Turkey (Armijo, 2007).

While the connection with mobile learning in higher education may be tenuous and indirect, the rise of these regions and countries does suggest challenges to the established pedagogies of North America and Western Europe that have historically informed higher education globally and challenges to the authority, resources and disposition of the states and institutions that underwrite these.

The Middle Circle—Global Universities

This next circle contains the global higher education system, responding to the economic and technological pressures and opportunities described above. In very general terms, global higher education has been characterized as becoming more global, corporate, competitive and stratified (Albach et al., 2009; Vaira, 2004). Technology has been co-opted into these changes, as universities and colleges use online and distance learning approaches to compete in more distant markets, and technologies replace human pedagogic and administrative functions, amounting to the creeping industrialization of the universities' and colleges' core business (Clegg et al., 2003; Traxler & Lally, 2015).

Universities from the global North establishing campuses around the world, commercial online trainers gaining degree-awarding powers, and metrics-driven teaching and research are all part of a more competitive, corporate and managerial ethos among the major players in higher education (Deem et al., 2008). Again, the impact on mobile learning is likely slow, haphazard and undocumented. It might, however, push research away from open-ended questions of purely pedagogic interest and towards questions with a direct operational impact on institution-level metrics and indicators such as recruitment, student satisfaction, degree classification, retention/progression and first-destination employment (Alexander, 2000). There are perceptions, especially outside the research community, that mobile learning is obvious, because everyone has experience of mobiles and has opinions on education, and that theorizing is redundant, superfluous, old-fashioned or elitist. This has led, on the basis of management decisions rather than a relevant and robust evidence-base, to some large-scale implementations.

We can expect to see this trend continue especially when combined with the growth of policies based on BYOD, especially when the latter are marketed as expanding consumer choice rather than fiscal expediency. If these policies become a reality rather than an aspiration, it will involve a substantial shift in attitudes about institutional control, support and training, not only in relation to technology and resources but also in relation to learning and knowledge. Currently, content and intellectual property (IP), secured within the institutional virtual learning environment (VLE) are regarded as critical business assets, only accessed by authorized users, namely fee-paying students. A shift towards learners' own devices, along with the corresponding services, networks, habits and expectations, will undermine that business model and the jobs of teachers as currently conceived.

The Next Circle

This circle is composed of the professions, the curricula, the missions and the practices of the universities and colleges that constitute higher education. At a global level, as we have said, over recent years, this has been characterized, certainly in the developed world, by increased global competitiveness, increased corporatization, and greater diversity (or stratification) of the universities themselves and a variety of new and increased pressures on university faculty to reconceptualize their roles and reconsider their priorities.

An earlier decade, as higher education was opened up to a larger and larger number of institutions and as student numbers rose, had seen the increasing professionalization of teaching, perhaps in order to assert the status of those less research-active institutions and less research-active faculty (Lueddeke, 2003). This stimulated the interest in pedagogic research alongside the more established subject research and helped feed into mobile learning research, especially where it worked with higher education students, and with practicing lecturers and was perceived as improving local learning outcomes and student satisfaction. While the funding that underwrote this may have eventually dried up, and while most innovation failed to embed, the medium-term outcome was part of a generation of faculty with a passing familiarity with some aspects of mobile learning. It was, however, often the case that in practice mobile learning was thought of as accessing the institutional VLE while communicating and receiving library recall notices by SMS. Only a smaller privileged handful of students experienced the richness of augmented reality field trips and other more exciting aspects of mobile learning (El-Hussein & Cronje, 2010 Mcconatha et al., 2008).

Obviously, also, the curriculum continues to evolve. Pure subject content changes as a consequence of subject research. Many higher education institutions are under continued pressure to produce *job-ready* graduates (Kavanagh & Drennan, 2008; Kivunja, 2014; Wang et al., 2009) and this drives *soft skills* (Andrews & Higson, 2008), digital literacy and problem-based learning *inter alia* into the curriculum, giving varied opportunities to mobile learning enthusiasts. The pressure also drives the introduction of more authentic assessment, including on-the-job assessment using mobiles (Dearnley et al., 2009; Pimmer & Pachler, 2014), but ongoing awareness of how plagiarism can taint a university's brand and damage its business creates a continued barrier to moving mobile learning from the unassessed and optional margins of learning and teaching to the assessed and credit-bearing core of the curriculum.

Nearly the Inner Circle

Mobile learning in higher education exists and engages with a host of other educational technology trends and fashions. At the moment, these include the flipped classroom, MOOCs, learner analytics (Ferguson, 2012), BYOD, lecture capture (Owston et al., 2011), digital literacy and content curation, with plagiarism detection (Carroll, 2002), online submission, and each new generation of virtual learning environment as neighbours in the educational technology ecosystem. To some extent, the longer-term viability of mobile learning depends on how it is perceived as a component, a competitor or a complement to any of these other trends, vying for resources, capacity and change. Some are conceptual and theoretical—for example, the connectivism (Siemens, 2014) that underpinned the early MOOCs—some are operational and pragmatic—for example, lecture capture and plagiarism detection—and that makes a difference. Features and attributes that might appeal to innovators and early adopters might not appeal to regulators, managers and rank-and-file faculty; the latter sometimes prefer whatever seems to offer a pragmatism, a more efficient version of the status quo, while the former seek novelty and surprise, and conceptual transformation.

The flipped classroom (Arshad & Imran, 2013; Bishop & Verleger, 2013; Hoffman, 2014; Hughes, 2013) was originally an American conception and a response to various practical and pedagogic problems and challenges (principally those of encouraging autonomous active learning, discussions about BYOD, and increasing pressure on contact time) and a response to a specific infrastructure and technical environment (namely the widespread availability of cheap connectivity and personal, social and domestic digital technologies). While an underpinning theory of the flipped classroom may have emerged, the idea was nevertheless originally nationally and culturally specific, and so are the documented experiences, which come originally from a very specific setting, namely schools in America. The flipped classroom concept should not be interpreted merely as using technology, for example mobile technology, to displace the consumption of content out of the classroom. In fact, the principles of the flipped classroom can be expressed as the following quite profound questions or challenges:

- How can we optimize the face-to-face learning experience? What is it that students can only get face-to-face with lecturers?
- How can we optimize the ways in which students can learn from each other face-to-face? What is it that students can only get face-to-face with each other?
- How can we optimize the campus experience? What is it that students can only get by coming on campus?

And:

- How can digital technology, including obviously mobile digital technology, support addressing these challenges? And what does digital technology, including mobile, do most effectively or uniquely?

In universities, these kinds of questions are being discussed, though perhaps not very rigorously. They are linked indirectly to earlier discussions about *personal response systems* (PRS), namely discussions about promoting more active and engaged learning in formal education. While the first incarnation of PRS was delivered on dedicated devices, the so-called *clickers* or *zappers* for example in economics (Elliott, 2003) and many other disciplines—and thus an unsustainable expense—later incarnations use software or apps, such as Socrative, that run on learners' own devices and thus align with emergent BYOD policies. Lecture capture, another current trend, is a technical fix to the problem of student attendance, English language proficiency and the need for enhanced revision, but is often implemented on systems, such as Panopto, that encourage playback on mobiles, and thus aligned to BYOD policies.

These policies are increasingly visible and the relationship with mobile learning is fairly clear. Furthermore, they seem to dig mobile learning out of the hole of its own making, of unsustainability. At a more operational level, however, BYOD policies present several local or tactical problems: first, the technical and infrastructural challenges; second, the operational aspects; and third, the ethical and political issues (Traxler, 2016). Mobile technologies are obviously the devices in question, albeit rather vaguely defined, but mobile learning has largely developed independently of the BYOD debate and discussion.

Content curation (Antonio et al., 2012), sometimes called orchestration, is the concept of supporting learning with resources from wherever the best were found, not defaulting automatically to the treadmill of in-house content creation. If content, and its IP, represent a perceived business asset within their respective universities, then curation is the basis for a very different business model. It is kindred to the open learning movement, driven by motivation that is part pedagogic, part ideological, part practical. To work effectively, curation requires teachers (or the designers of learning experiences) to develop skills around the selection and organization of resources. Given that the mobile is the portal to the resources of the Web for an almost universal demographic beyond the classes and institutions that access networked computers, there is a clear and powerful connection between mobile learning and content curation, but higher education is still largely committed to in-house resources. Skills and attitudes are the likely barrier to change, though it is perfectly reasonable to argue that in a few cases in a few institutions in-house content encapsulates unique new knowledge, unavailable anywhere external.

The movement to exploit external content, that is the curation of content, is aligned to the movement to exploit user-generated or learner-generated content (Lee & McLoughlin, 2007). This is not necessarily a mobile phenomenon, but the mobile empowers a more broadly based demographic of creators and richer and diverse content, derived from the environment and the context, wherever people can take their mobiles. It implicitly happens every time someone shares their ideas and images on social media (and similarly, curation happens every time someone uses social book-marking), but it is only a fringe activity in higher education.

Digital literacy is the pervasive and comprehensive underpinning of all these various trends and formulations. In its broadest interpretation, digital literacy describes those skills, attitudes, access and competences necessary for individuals, and perhaps communities, to flourish in an increasingly digital world (Bawden, 2008). The genesis of digital literacy was higher education in Western Europe, with a clearly defined pressure to focus on elearning, IT skills and employability and a less coherent lobby for the social, culture, expressive and creative aspects of being digital (Martin, 2005). A specifically mobile element of digital literacy has yet to be developed in spite of the observation about how pervasive and ubiquitous mobiles are in our societies. This may illustrate the tension between the educational technologies of the institutions as opposed to the universal technologies of their societies.

A discussion of the MOOC should follow logically from all the previous topics (Baggaley, 2013; Johnson et al., 2014; Ng & Widom, 2014). The idea of the MOOC was born out of experiences with large open online courses in higher education that suggested a new pedagogy, where the numbers and connections would create a new learning paradigm. The story of the MOOC is, however, not straightforward. In its original formulation, it did indeed hark back to many of the issues just discussed. It grew out of ideas and experiences with large numbers of connected learners, termed *connectivism*. As well as the well-documented Canadian experiments, there were small ones with a specifically free and mobile focus (de Waard et al., 2011). The idea of the MOOC became, however, co-opted by formal institutional perspectives and now has been transformed into a highly interactive rich media experience broadcast by universities on a small number of specialized and dedicated platforms (Kop & Hill, 2008). Some platforms more than others render well on mobile devices—FutureLearn in the UK is a prime example of a *mobile first* philosophy (Sharples et al., 2015)—but the early idealism of the *wisdom of the crowd* (Halasek et al., 2014) has been replaced by a globally competitive and corporate ethos.

Alongside the issues of the changes, fashions and trends that constitute the immediate environment of mobile learning in higher education we must also recognize elements of stasis and stagnation; whatever the rhetoric, the learning management system (LMS) or VLE has become the established repository of institutional content and change is often only represented by periodic replacement of one model or brand by its obvious successor rather than by any more fundamental debate. In this sense, VLE or LMS have become an established part of the university experience or the university tradition, of which learner analytics, plagiarism detection and online submission are just minor appendages or refinements. Consequently, in this depiction, mobile learning may only mean accessing the VLE or receiving SMS alerts on a mobile, and innovation is necessarily severely constrained at an institutional level; the enacted version of mobile learning in higher education may be quite different from the espoused version enunciated by its early fathers.

The Inner Circle

This circle is probably the comfort zone of the mobile learning research community, where the challenges are to some extent contrived and controlled, where the research agenda is the one we, as a research community, have chosen and perfected. Research journals continue to publish individual accounts (for example, the *Journal of Mobile and Blended Learning*, the *British Journal of Educational Technology*, *Research in Learning Technology*, *Computer and Education* and many more) and edited books continue to consolidate these. Global, national and regional conferences promote works in progress: for example, the annual *mLearn* and the research body, the *International Association for Mobile Learning*, nearing its tenth year, create a network and a forum. This is a small, stable and secure ecosystem and, as the current book underlines, its challenge is reaching out across its boundaries rather than surviving within them.

In other words, one aspect of the mobile learning research community within higher education was the extent to which it bought into the rhetoric and funding associated with *innovation*; the community portrayed itself as *innovative*, or even *disruptive*, partly because in relation to the established institutional elearning research, which already bought into the same rhetoric, this was undoubtedly accurate.

This was, it turns out, wrong for several different reasons (Traxler, 2011). One, already mentioned, was the drawing back of the state intervention that funded innovation and the decrease in global economic resources to provide the funding. Another was the failure generally of innovation to use evidence to unlock sustained funding, to move from the institutional periphery to the institutional centre, and a related reason was the tendency in early pilots to use a uniform set of devices issued to the participants, producing rigorous but irrelevant results that told universities nothing about the sustainable diversity of user-owned devices. Related reasons included the necessary emphasis on hardware and infrastructure in the early days when these were challenging, telling universities nothing in later days when the hardware and infrastructure in question had become robust and irrelevant. So, the pilots of the past taught us little about the problems of the present. As we said earlier, even within higher education, champions and advocates of mobile learning must react and respond to the challenges and opportunities presented by the flipped classroom, the BYOD movement, lecture-capture, user-generated content, digital literacy and content curation (or orchestration), but a more conservative consumerist environment with little appetite or resources for innovation or risk means any of these topics remains at the margins and they must also work with a perception in many universities that mobile learning is merely Moodle on a smart-phone, library recalls as text messages and, for the more affluent institutions, free tablets for the next few intakes of students studying medicine or geography (challenged in Cochrane et al., 2016).

There is, however, a deeper tension, one between mobile learning in universities as opposed to universities in mobile and connected societies. As we have hinted, the inward-looking and backward-looking trajectory of mobile learning as originally conceived has run its course, but universities have not come to grips, except in a haphazard, piecemeal and tactical way, with the implications of the changed world around them and how it challenges them to think about their relevance, value, authority and credibility.

Concluding the Contexts and the Challenges

The conclusion must be that there are no conclusions; there is too little evidence, too much context, too much change and too little time, movement is rapid, progress uncertain. This book is a snapshot, capturing the diverse enthusiasms and achievements of our contributors, assuring us that we are all building a more engaging and attractive conception of higher education even if we cannot quite discern its shape or composition.

References

Alexander, F. K. (2000). The changing face of accountability: Monitoring and assessing institutional performance in higher education. *The Journal of Higher Education*, *71*(4), 411–431.

Altbach, P. G., Reisberg, L., & Rumbley, L. E. (2009). Trends in global higher education: Tracking an academic revolution. *A Report Prepared for the UNESCO 2009 World Conference on Higher Education*. Geneva: United Nations Educational, Scientific and Cultural Organization.

Andrews, J., & Higson, H. (2008). Graduate employability, 'soft skills' versus 'hard'business knowledge: A European study. Higher education in Europe, *33*(4), 411–422.

Antonio, A., Martin, N., & Stagg, A. (2012). Engaging higher education students via digital curation. In *Future challenges, sustainable futures. Proceedings of the 29th Australasian Society for Computers in Tertiary Education Conference* (pp. 55-59). 25-28 Nov 2012, Wellington, New Zealand.

Armijo, L. E. (2007). The BRICs countries (Brazil, Russia, India, and China) as analytical category: mirage or insight? *Asian Perspective*, *32*(4), 7–42.

Arshad, K., & Imran, M. A. (2013). Increasing the interaction time in a lecture by integrating flipped classroom and just-in-time teaching concepts. *Compass: Journal of Learning and Teaching*, *4*(7).

Baggaley, J. (2013). MOOC rampant. *Distance Education*, *34*(3), 368–378.

Bawden, D. (2008). Origins and concepts of digital literacy. In C. Lankshear & M. Knobel (Eds.), *Digital literacies: Concepts, policies and practices* (pp. 17–32). New York: Peter Lang.

Bishop, J. L., & Verleger, M. A. (2013). The flipped classroom: A survey of the research. In *ASEE National Conference Proceedings*, Atlanta, GA.

Carroll, J. (2002). Editors' Introduction. In *A handbook for deterring plagiarism in higher education* (p. 119). Oxford, UK: OCSLD.

Clegg, S., Hudson, A., & Steel, J. (2003). The Emperor's new clothes: globalisation and e-learning in higher education. *British Journal of Sociology of Education*, *24*(1), 39–53.

Cochrane, T., Jones, S., Kearney, M., Farley, H. & Narayan, V. (2016). Beyond Pokémon Go: Mobile AR & VR in education. In S. Barker, S. Dawson, A. Pardo, & C. Colvin (Eds.), *Show Me The Learning. Proceedings ASCILITE 2016*, Adelaide (pp. 136-138).

Dearnley, C., Taylor, J., Hennessy, S., Parks, M., Coates, C., Haigh, J., Fairhall, J., Riley, K., & Dransfield, M. (2009). Using mobile technologies for assessment and learning in practice settings: Outcomes of five case studies. *International Journal on E-Learning*, *8*(2), 193–207.

Deem, R., Mok, K. H., & Lucas, L. (2008). Transforming higher education in whose image? Exploring the concept of the 'world-class' university in Europe and Asia. *Higher Education Policy*, *21*(1), 83–97.

Demyanyk, Y., & Van Hemert, O. (2011). Understanding the subprime mortgage crisis. *Review of Financial Studies*, *24*(6), 1848–1880.

De Waard, I., Abajian, S., Gallagher, M. S., Hogue, R., Keskin, N., Koutropoulos, A., & Rodriguez, O. C. (2011). Using mlearning and MOOCs to understand chaos, emergence, and complexity in education. *The International Review of Research in Open and Distributed Learning, 12*(7), 94–115.

El-Hussein, M. O. M., & Cronje, J. C. (2010). Defining mobile learning in the higher education landscape. *Educational Technology & Society, 13*(3), 12–21.

Elgort, I. (2005, December). E-learning adoption: Bridging the chasm. In Ascilite (pp. 181–185). Retrieved from http://www.ascilite.org/conferences/brisbane05/blogs/proceedings/20_Elgort.pdf

Elliott, C. (2003). Using a personal response system in economics teaching. *International Review of Economics Education, 1*(1), 80–86.

Ellis, R., & Goodyear, P. (2013). *Students' experiences of e-learning in higher education: the ecology of sustainable innovation*. New York: Routledge.

Ferguson, R. (2012). Learning analytics: drivers, developments and challenges. *International Journal of Technology Enhanced Learning, 4*(5–6), 304–317.

Furfie, B. (2010). Is the iPad a game changer? *Engineering & Technology, 5*(4), 34–35.

Halasek, K., McCorkle, B., Selfe, C. L., DeWitt, S. L., Delagrange, S., Michaels, J., & Clinnin, K. (2014). A MOOC with a view: How MOOCs encourage us to reexamine pedagogical doxa. In S. Krause & C. Lowe (Eds.), *Invasion of the MOOCs: The promise and perils of massive open online courses* (pp. 156-166). Anderson, SC: Parlor Press.

Herrington, A., & Herrington, J. (2007). Authentic mobile learning in higher education. In P. Jeffrey (Ed.), Proceedings of the Australian Association for Research in Education (AARE) International Educational Research Conference (pp. 1-9). Australia: AARE.

Herrington, J., Herrington, A., Mantei, J., Olney, I. W., & Ferry, B. (2009). New technologies, new pedagogies: Mobile learning in higher education. In R. Atkinson, & C. McBeath (Eds.), Annual Conference of the Australasian Society for Computers in Learning in Tertiary Education, p. 419-427, Melbourne, Australia: Deakin University.

Hill, D., & Kumar, R. (Eds.), (2012). Global neoliberalism and education and its consequences. New York: Routledge.

Hoffman, E. S. (2014). Beyond the flipped classroom: Redesigning a research methods course for e^3 instruction. *Contemporary Issues in Education Research*, 7(1), 51–62.

Hughes, H. (2013). Flipping the college classroom: Participatory learning. In L. Kyei-Blankson, & E. Ntuli (Eds.), *Practical Applications and Experiences in K–20 Blended Learning Environments* (pp. 137–152). Hershey, PA: Information Science Reference.

Johnson, D., Nafukho, F., Valentin, M., Lecounte, J., & Valentin, C. (2014). The origins of MOOCs: The beginning of the revolution of all at once-ness. In Proceedings of 15th International Conference on Human Resource Development research and practice across Europe. Edinburgh, UK: Edinburgh Napier University Business School.

Kavanagh, M. H., & Drennan, L. (2008). What skills and attributes does an accounting graduate need? Evidence from student perceptions and employer expectations. *Accounting & Finance, 48*(2), 279–300.

Kivunja, C. (2014). Do you want your students to be job-ready with 21st century skills? Change pedagogies: a pedagogical paradigm shift from Vygotskyian social constructivism to critical thinking, problem solving and Siemens' digital connectivism. *International Journal of Higher Education, 3*(3), 81.

Kop, R., & Hill, A. (2008). Connectivism: Learning theory of the future or vestige of the past? *The International Review of Research in Open and Distributed Learning, 9*(3).

Kumar, S. (2004). Mobile communications: global trends in the 21st century. *International Journal of Mobile Communications, 2*(1), 67–86.

Lakes, R. D., & Carter, P. A. (2011). Neoliberalism and education: An introduction. *Educational Studies, 47*(2), 107–110.

Lee, M. J., & McLoughlin, C. (2007). Teaching and learning in the Web 2.0 era: Empowering students through learner-generated content. *International Journal of Instructional Technology and Distance Learning*, *4*(10), 21–34.

Lee, Y. H., Hsieh, Y. C., & Hsu, C. N. (2011). Adding innovation diffusion theory to the technology acceptance model: Supporting employees' intentions to use e-learning systems. *Educational Technology & Society*, *14*(4), 124–137.

Lueddeke, G. R. (2003). Professionalising teaching practice in higher education: A study of disciplinary variation and 'teaching-scholarship'. *Studies in Higher Education*, *28*(2), 213–228.

Mcconatha, D., Praul, M., & Lynch, M. J. (2008). Mobile learning in higher education: An empirical assessment of a new educational tool. *TOJET: The Turkish Online Journal of Educational Technology*, 7(3).

Martin, A. (2005). DigEuLit–a European framework for digital literacy: a progress report. *Journal of eLiteracy*, *2*(2), 130–136.

Ng, A., & Widom, J. (2014). Origins of the Modern MOOC (xMOOC). Hrsg. Fiona M. Hollands, Devayani Tirthali: MOOCs: Expectations and Reality: Full Report, 34–47.

Ng, W., & Cumming, T. M. (Eds.) (2015). *Sustaining mobile learning: Theory, research and practice*. New York: Routledge.

Olssen, M., & Peters, M. A. (2005). Neoliberalism, higher education and the knowledge economy: From the free market to knowledge capitalism. *Journal of Education Policy*, *20*(3), 313–345.

Owston, R., Lupshenyuk, D., & Wideman, H. (2011). Lecture capture in large undergraduate classes: Student perceptions and academic performance. *The Internet and Higher Education*, *14*(4), 262–268.

Pimmer, C., & Pachler, N. (2014). Mobile learning in the workplace: Unlocking the value of mobile technology for work-based education. In M. Ally, & A. Tsinakos (Eds.), *Increasing access through mobile learning* (pp. 193-203). Athabasca, Canada: Athabasca University.

Romiszowski, A. J. (2004). How's the e-learning baby? Factors leading to success or failure of an educational technology innovation. *Educational Technology*, *44*(1), 5–27.

Salmon, G. (2005). Flying not flapping: a strategic framework for e-learning and pedagogical innovation in higher education institutions. *ALT-J*, *13*(3), 201–218.

Sharples, M., Kloos, C. D., Dimitriadis, Y., Garlatti, S., & Specht, M. (2015). Mobile and accessible learning for MOOCs. *Journal of Interactive Media in Education*, *2015*(1). doi: http://doi.org/10.5334/jime.ai

Siemens, G. (2014). *Connectivism: A learning theory for the digital age. International Journal of Instructional Technology and Distance Learning*, *2*(1), 3-10.

Traxler, J. (2011) Mobile learning - Starting in the right place, going in the right direction? *International Journal of Mobile and Blended Learning*, *3*(1), 57–67.

Traxler, J. (2016) Inclusion in an age of mobility, *Research in Learning Technology*, *24*.

Traxler, J., & Lally, V. (2015) The crisis and the response: After the dust had settled. *Interactive Learning Environments*, *24*(5), 935–937.

Vaira, M. (2004). Globalization and higher education organizational change: A framework for analysis. *Higher Education*, *48*(4), 483–510.

Wang, J., Ayres, H., & Huyton, J. (2009). Job ready graduates: A tourism industry perspective. *Journal of Hospitality and Tourism Management*, *16*(1), 62–72.

Wortham, J. (2009). Apple's game changer, downloading now. *The New York Times*, New York, NY, BU1.

INDEX

Note: page numbers in bold are cross-referencing terms found in tables; page numbers in italics are cross-referencing terms found in figures.